SHOUT-OU
THE LNT

"With enough facts and juicy historical details to satisfy most of us, The LNT Report *reads like a detective story, sprinkled with (nerdy) humor here and there. Layer by layer, it uncovers the BS (Bad Science) behind much of our nuclear fear."*

> —RAULI PARTANEN, author of *The Age of Energy* and *Climate Gamble*

"This exposé, with wit and clarity, debunks the nonsense that has prevented deployment of our safest energy source with the smallest environmental footprint—modern nuclear power. For the sake of young people, this story must be spread widely."

> —PROFESSOR JAMES HANSEN, Earth Institute, Columbia University

"The author has dug out the story, something that the EPA and the scientific community have failed to do . . . One can only hope that The LNT Report *will be read by all students from high school on, in this country and around the world."*

> —PROFESSOR EDWARD J. CALABRESE, University of Massachusetts Amherst

"Conley provides a blow-by-blow account of how bad science and outright fraud became the basis of our regulations on radiation safety."

> —JOSHUA GOLDSTEIN, author of *A Bright Future* and co-writer of Oliver Stone's famous documentary movie, *Nuclear Now*

"Public attitudes on the risks of nuclear energy are dominated by mythical thinking, not scientific evidence. This history of scientific work on the effects of radiation deserves a wide audience and careful consideration."

> —SPENCER WEART, author of *The Rise of Nuclear Fear*

"Ed Calabrese is a brilliant scientific detective, and Mike Conley is a brilliant scientific explainer. No ifs or buts, if you have any rational faculty whatsoever, this book will absolutely convince you. The dangers of nuclear energy have been outrageously exaggerated!"

> —DR. RAY SCOTT PERCIVAL, creator of *Enlightenment Defended*, author of *The Myth of the Closed Mind*

"Fear of Nuclear Energy is far more dangerous than nuclear energy itself."

> —ROBERT BRYCE, author of *A Question of Power*

"If you ask, Does LNT make sense?, Conley's step by step historical narrative with citations, for better or for worse, explains it all."

> —RAY ROTHROCK, nuclear engineer, venture capitalist, philanthropist

THE
LNT REPORT

Other books on Nuclear Energy from Open Universe

Earth Is a Nuclear Planet: The Environmental Case for Nuclear Power, by Mike Conley and Tim Maloney, 2024

The Poverty of Green Philosophy: A Marxist Case for Nuclear Energy in a Cooperative World, by Bill Sacks and Greg Meyerson, 2025

Roadmap to Nowhere: A Reality Check on Renewable Energy by Mike Conley and Tim Maloney, 2026

THE
LNT REPORT

HOW BAD SCIENCE
MADE THE WORLD AFRAID
OF NUCLEAR POWER

BY

MIKE CONLEY

MANUSCRIPT EDITED BY

TIM MALONEY

OPEN UNIVERSE
Chicago

To find out more about Open Universe and Carus Books, visit our website at www.carusbooks.com.

First edition 2025

Published by arrangement with Generation Atomic, GenerationAtomic.org.

Cover art Aaftab Sheikh, www.Instagram.com/aaft.dedsigns. Adapted by Shane Arbogast.

Illustrator: Erwan Sulistyo, erwan.syah21@gmail.com, https://www.fiverr.com/erwand

Printed and bound in the United States of America. Printed on acid-free paper.

The LNT Report: How Bad Science Made the World Afraid of Nuclear Power

ISBN: 978-1-63770-065-5

This book is also available as an e-book (978-1-63770-066-2).

Library of Congress Control Number: 2024937353

Reviewed by:

Dr. Ed Calabrese, Ph.D., Professor of Toxicology and Risk Management, University of Massachusetts at Amherst

Senior Science Advisor:

Stephen A. Boyd, Ph.D., Chemical Physics / Solid-State Chemistry, Stony Brook

Science Editors:

Robin Salter, Ph.D., Genetics, University of Wisconsin at Madison, Professor Emeritus of Biology, Oberlin College

Leonard Rodberg, Ph.D., Nuclear Physics, Massachusetts Institute of Technology, Professor Emeritus of Urban Studies, Queens College / CUNY, Research Director of Nuclear New York

Philip Hult, M. Phil., Nuclear Engineering, University of Cambridge

Proofread by:

Tim Maloney, Scott Medwid

Dedication

This book is dedicated to Dr. Edward Calabrese, professor of toxicology and risk management at the University of Massachusetts at Amherst. His tireless pushback against a century of disinformation and fear mongering about low-dose ionizing radiation is an inspiration to scientists and citizens alike.

In his nearly fifty years at Amherst, Calabrese has researched and taught the subject of dose response to toxins, radiation, and other stressors, writing several books and more than 500 papers on the subject.

And then his work took an interesting turn. From his 2018 Congressional testimony:

> More recently in the past decade, I have exhaustively researched the historical origins and scientific basis of EPA's LNT model [the Environmental Protection Agency's Linear No-Threshold model of radiation safety] and found it sorely wanting. LNT is important because it is the model upon which all our cancer risk assessment and key health and ecological regulations are based.
>
> What I have learned was unexpected, and has turned more than thirty years of my understanding of toxicology upside down. It has revealed that what I had taught for so many years at U Mass and written about so ardently in my many articles and books was factually wrong.

What began for me as a routine academic exercise to affirm the scientific origins and credibility of LNT ironically ended as a remarkable repudiation of its scientific adequacy.[1]

This plain-language introduction to Dr. Calabrese's research on the Linear No-Threshold hypothesis—that there is no safe dose of radiation, and that all doses are cumulative—is offered as a tribute to the growing impact of his work. Our hope is that the history of bad science he has painstakingly compiled will mainstream the critical facts about low-dose radiation, and dissipate the fog of nuclear fear that has pervaded the issue of clean energy production for the last several decades.

Ed Calabrese is a hero in the nuclear power community. After reading this short volume, we're confident you will understand why.

Contents

Introduction

THIS IS A TRUE-CRIME STORY. Not a whodunnit, where the basic facts are known while the perpetrator remains a mystery, but a case like Lieutenant Columbo would tackle. We see the crime committed right up front, then watch how Columbo unmasks the perpetrator, discredits their lies, and brings them to justice. As the story unfolds, we discover the sordid details of how the crime was conceived, executed, and covered up, and learn how the perpetrator's fatal flaws led to their undoing.

The crime in question is the attempted (and nearly successful) murder of nuclear power through Bad Science, ginned up with assumptions and contrary facts while lacking any solid evidence. But unlike a *Columbo* episode, we're left with a cliffhanger: Now that the mistakes, cover-ups, and misinformation have been tracked down and exposed, will there still be time to save humanity from the looming threat of climate change?

This is the long, strange tale of LNT ('Linear No-Threshold'), the late Dr. Hermann Muller's claim that any amount of nuclear radiation, even a tiny dose, does some amount of harm. Devised nearly a century ago, and still considered gospel by far too many people, the LNT hypothesis falsely claims that there is no safe dose of radiation and that all doses are cumulative. As the story unfolds, you'll discover how the regulations for low-dose ionizing radiation have been based on bad science (BS) right from the start, and you'll learn what can be done to remedy the error.

After working with Ralph Nader on environmental issues in the mid-seventies, Ed Calabrese became a professor of toxicology and risk assessment at the University of Massachusetts at Amherst, where he still works today. Among his many accomplishments, he won the 2009 Marie Curie Award for his groundbreaking work on hormesis (not homeopathy, but hormesis—see Chapter 11). Over the last several years, Calabrese has been a scientific Columbo, carefully documenting how and why the LNT model became the world's radiation safety standard, even though:

- Muller's hugely influential 1927 paper was never peer-reviewed.

- He never published an experiment for others to replicate his work or verify his claims.

- His LNT model has been roundly refuted by modern science.

Establishing LNT as the default model of radiation risk assessment required the sustained action of influential and deep-pocket supporters:

- The petroleum-rich Rockefeller Foundation funded the original research, and influenced the 1956 National Academy of Sciences (NAS) to adopt the LNT model as public health policy.

- Detlev Bronk, the president of the NAS at the time, was also the president of the Rockefeller Institute of Medical Sciences (later Rockefeller University).

- The June 1956 NAS pamphlet *Report to the Public* on LNT was widely promoted and distributed by the *New York Times*, whose chairman Arthur

Sulzberger was on the board of the Rockefeller Foundation.

- Since that time, a growing body of evidence contrary to LNT has been systematically suppressed, ignored, or shouted down.

Since the advent of commercial nuclear power in the 1950s, Hermann Muller's fatally-flawed "no-safe-dose" hypothesis on radiation risk assessment has frustrated humanity's efforts to clean up our energy act. For nearly a hundred years, Muller's incorrect assertions have conjured a gnawing sense of anxiety in the public psyche, slowing the buildout of nuclear power, the world's most reliable clean-energy solution, and the safest one as well.

More than a fact-finding investigation, the historical research of Dr. Calabrese has brought to light a tragic tale of brilliant minds gripped by pride, power, and blind ambition, Shakespearean flaws that have complicated so many endeavors down through the ages. But as you learn who the players are and what they did, keep in mind that they were not cardboard characters with ill intent, but highly competitive and all-too-human explorers trying to make sense of the world around them. Working at the bleeding edge of knowledge, with the primitive equipment available at the time, and tantalized by the possibility of world-changing discoveries, they sometimes fell short of the humility, dispassionate analysis, and rigorous standards demanded by the Scientific Method. This is a moral lesson about why facts matter.

Whatever their intentions may have been, it is incumbent upon the scientific community to acknowledge the errors and correct the mistakes, and incumbent upon governments around the world to update their rules and regulations accordingly. At the same time, it is equally incumbent upon us, the curious public, to never forget that we too have all made assumptions we have stridently defended, long after the evidence has turned against us.

This, then, is a story of people, not monsters, but a true-crime story nonetheless.

The body of evidence assembled by Dr. Calabrese was the principal source for this book; the titles of some of his papers will give you a sneak preview of the chapters ahead. Take a moment to scan the list and you'll see what we mean.[1] Also recommended is his series of twenty-two interviews by the Health Physics Society,[2] portions of which are linked to at several points as the story unfolds.

More than any other factor, Muller's LNT model of radiation risk assessment has been the principal bottleneck restricting the advance of nuclear power. Muller's false ideas, codified a half-century ago into our regulatory standards and practices, have needlessly driven up the cost of commercial nuclear power in a misguided effort to prevent the release of the most negligible wisps of radiation, while costing a needless fortune. Like the claim of a stolen election with no evidence to back it up, LNT is the 'big lie' in antinuclear circles, an enduring article of faith that has undermined the conversation on clean energy for nearly seventy years.

Although this is a stand-alone volume, designed for readers with no prior knowledge of the topic, it also serves as a companion piece to the larger work by Tim Maloney and myself, *Earth Is a Nuclear Planet*. As nuclear energy advocates, it has become clear to us that Hermann J. Muller's Linear No-Threshold (LNT) model of radiation risk assessment, vintage 1927, is the root of the world's unwarranted nuclear fear. So, while Tim was on a well-deserved hiatus, I wrote this little book.

Our earlier work *Earth Is a Nuclear Planet: The Environmental Case for Nuclear Power* explores nuclear energy, radiation, and nuclear safety in detail. Its upcoming sequel *Roadmap to Nowhere: A Reality Check on Renewable Energy* compares a popular proposal for a 100-percent national renewable-energy grid with a 100-percent nuclear grid (spoiler alert—nuclear wins hands down).

MAKE ENERGY

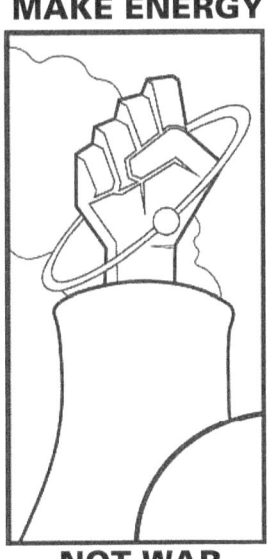

NOT WAR

Our books address this key question:

How do we power a planet of eight billion people without de-spoiling the natural world?

In our view, the best solution with the smallest footprint would be a rapid expansion of nuclear power, but Muller's flawed risk model has complicated the buildout of this vital solution to the world's energy needs. His mistaken ideas have been the basis of the world's nuclear safety regulations since the 1950s.

Rather than increasing public safety and confidence, LNT-based rules and regs have increased the public's nuclear fear, dissuading too many of us from the best technology we have to address the climate crisis by providing the abundant clean power we need both now and in the years to come.

Simply put—the LNT bottleneck must be removed for humanity to flourish and prosper. And that requires a clear understanding of why the LNT is bad science.

CHAPTER ONE
Hermann Muller (1927–1932)

IN THE 1920S, the health effects of low dose radiation were a scientific afterthought. In the field of genetics, for example, researchers were trying to induce inherited mutations by zapping laboratory specimens with high doses of radiation. The race was on to produce genetic changes at will; it was believed that discovering the mechanism of inherited mutations would unlock the secrets of evolution. Whoever got there first would be a shoo-in for the Nobel Prize.

The competition was fierce, and on July 22nd 1927 the journal *Science* published Hermann Muller's landmark paper "The Artificial Transmutation of the Gene," in which he claimed to have induced inherited mutations at will in the offspring of *Drosophila* fruit flies.[1] The genetics community was thrilled—someone had finally done it!

Muller had been trying to induce gene mutation since graduate school. In the mid-1920s, under the guidance of famed geneticist Thomas Hunt Morgan at Columbia University in New York City,[2] Muller and his fellow doctoral candidates had taken a kitchen-sink approach to the problem, throwing everything they could think of at their hapless fruit-fly specimens. Muller's preferred method was to blast them with higher and higher doses of ionizing radiation, the type of radiation sufficiently energetic to knock an electron away from an atom. This turns the atom into an ion of its former self, which alters the chemical composition of whatever molecule the atom is a part of. And if it's part of a chromosome molecule in a living cell, genetic consequences can result. With the high doses he was using, most of his flies became sterilized and a lot of them died, but he mated the ones that survived to see what he could see.[3]

A few years later, as a professor at the University of Texas at Austin, Muller shot to prominence in the summer of 1927 with what many thought was the first published claim of tangible results. Actually, Stuart Gager and Albert Blakeslee had beaten him by six months with a January 1927 paper in *Proceedings*, the journal of the US National Academy of Sciences, in which they described using radiation to induce inherited mutations in plants. Muller failed to cite them in his paper and in his subsequent studies as well, even though Gager and Blakeslee kept reminding him in their own papers that they were the first to publish.[4] Despite their seminal efforts, it was the publication of Muller's July 1927 paper that attracted all the attention, and he was invited to present his findings at the Fifth International Genetics Conference in Berlin that September, just a few weeks away.[5] His speech electrified the audience, securing his position as the new trailblazer.

Less than a year after the confab in Berlin, Gilbert Lewis,[6] the renowned chemist at UC Berkeley, famous for developing the science behind Lewis acids and bases, came across Muller's paper. Although Lewis and his colleague Axel Olson were chemists and not geneticists, they had been speculating that mutations caused by continuous, low-dose cosmic and terrestrial "background" radiation, something that all species encounter, may hold the key to evolution.[7] It was their articulation of the Linear No-Threshold (LNT) concept that brought wider attention to Muller's work, giving it a name and introducing the idea to a global audience of scientists extending far beyond the genetics community.

Stepping out of their field of chemistry and into the realm of biology, Lewis and Olson had been ruminating along these lines as early as 1925.[8] And now, here was this young geneticist providing the confirmation they were seeking. Muller's short blasts of ultra-high-dose radiation did indeed induce mutation, so Lewis and Olson reasoned that if all doses were cumulative, as Muller claimed them to be, then perhaps the continuous low-dose background radiation that every life form receives on this nuclear planet of ours might have the same effect, causing the inherited mutations that result in evolutionary change.

While the 1927 Berlin conference made Muller famous in the world of genetics, Lewis and Olson's 1928 paper gave him a much wider scientific celebrity. Even so, Muller argued in a 1930 paper against their speculation that low-dose background radiation may be the source of hereditary mutations—the toast of the ball wasn't dancing with thems that brung him.[9]

In the midst of his growing celebrity, some of Muller's colleagues in the US, including his best friend and pen pal Dr. Edgar Altenburg,[10] were having doubts about his work. An accomplished radiation geneticist in his own right, Altenburg argued that Muller was ignoring an equally plausible explanation. Instead of inducing "point mutations" in the chromosomes of his fruit flies as Muller claimed, he may have been causing massive gene deletions instead, by blasting big holes in their chromosomes and destroying entire sections of genetic code.[11] His strange-looking *Drosophila* offspring may have been the consequence of using what amounted to a radiation bazooka on their parents, with doses at a whopping 130 million times greater than the average background dose.[12]

Muller strongly disagreed, and at the time there was no way to prove either of them right or wrong. Meanwhile at UT Austin, Professor Muller was having his students run more *Drosophila* experiments, this time using four doses that ranged from "only" 25 to 30 million times background radiation—a fraction of his original research range.[13] In Muller's view, a proportional response at these lower doses (which were still

3

fantastically high) would satisfy the skeptics and quash the argument.

As Muller predicted, his students found a linear, or proportional, response at these lower dose rates. Because of their still exorbitantly high range, the doses produced a "linear" effect—the more radiation, the more mutations. In one of his Health Physics Society interviews,[14] Dr. Calabrese describes what happened next (lightly edited for clarity):

> Muller had two students who did a study each, and I'll call them good studies for the time period [1930]. He guided them, and [their studies] showed a linear dose-response relationship; I wouldn't dispute that at all. It's exactly what the data said—it showed a linear dose response in the range that was tested. But what Muller then did was he decided to extrapolate.
>
> He went outside the observed data, but he didn't just go kind of outside the observed data by a little bit. [His extrapolation] went from about 130 million times above background [dose], down to background [dose] and below [i.e., to zero dose].
>
> It was kind of like saying, well, I know what the weather is today and maybe for the next couple of days, so I'm going to extrapolate on what the weather forecast will be three years from now. In effect, that's what Muller was doing. He was going way beyond his data and extrapolating into a vast, unknown zone.

With no solid science to back him up, Muller taught his students to extend the straight-line response they saw at high doses into the untested (and to this day still untestable) low-dose range, all the way down to background dose. This was before Muller was even using the term linear no-threshold. At the time, he called it his Proportionality Rule.[15]

In the 1920s, Gregor Mendel's pea plant experiments from the 1860s were still the state of the art in genetics research.[16] No one had yet been able to directly observe the DNA molecule in any detail, much less the tiny base-pairs (A-T and C-G) that form the ladder rungs of its double-helix configuration. Sixty

FIGURE 1: Muller's 1929 LNT Fruit-Fly Experiments

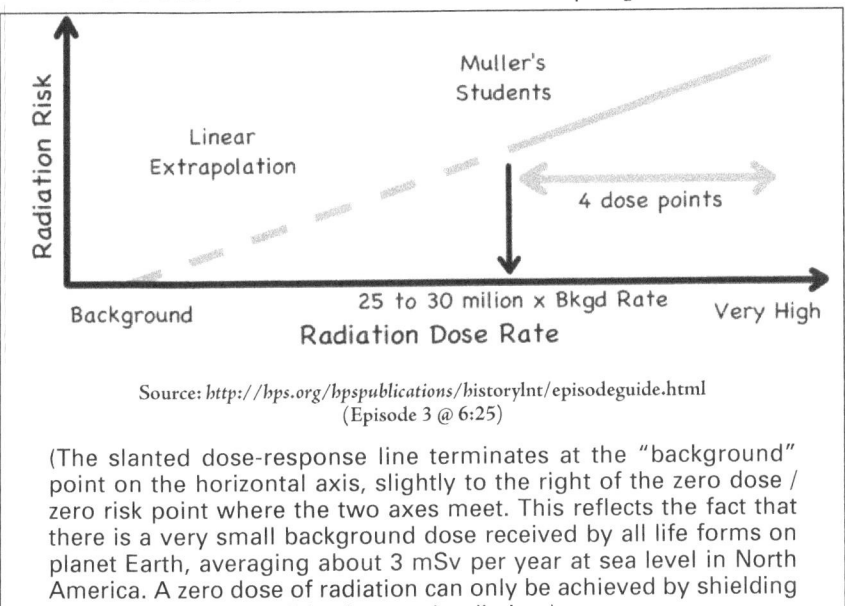

Source: *http://hps.org/hpspublications/historylnt/episodeguide.html*
(Episode 3 @ 6:25)

(The slanted dose-response line terminates at the "background" point on the horizontal axis, slightly to the right of the zero dose / zero risk point where the two axes meet. This reflects the fact that there is a very small background dose received by all life forms on planet Earth, averaging about 3 mSv per year at sea level in North America. A zero dose of radiation can only be achieved by shielding lab specimens from all background radiation.)

FIGURE 2: Basic Structure of the DNA Molecule with
A-T and G-C pairs

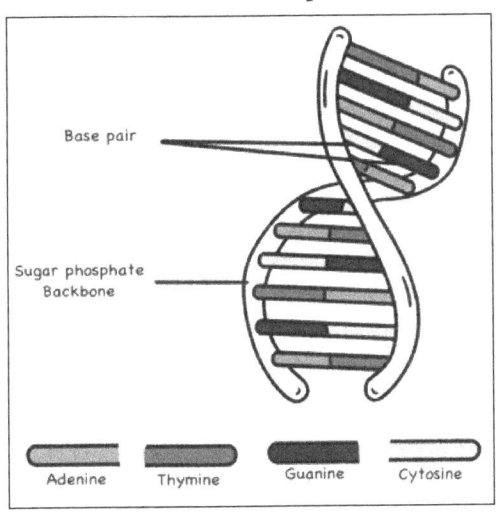

years after Mendel's work, geneticists were still limited to observing physical changes in a specimen (eye color, antenna length, wing size, and so forth) and hazarding a guess as to what may have changed in the parents' genes to trigger these inherited mutations.

By the 1920s, genes were already known to be the blueprint for all living things, from single-cell organisms to the largest mammals, and it was known that that the details were somehow encoded in the enormous chromosome molecule. The problem was, no one could see the blueprint. The technology did not yet exist to directly observe the molecule itself, and thus prove or disprove Muller's claim.

That was about to change.

CHAPTER TWO
Barbara McClintock and Lewis Stadler (1931–1954); Warren Spencer (1944–1947)

IN THE EARLY 1930S, Barbara McClintock, a corn cytoge-
neticist at Cornell University, discovered how to stain the inte-
rior of cells.[1] Her technique exposed cellular structures for the
first time to direct observation with existing microscopes. Some
fifty years later, she would be awarded the Nobel Prize for her
work on "jumping genes"—DNA sequences that physically move
from one position on the chromosome to another. But back in
the thirties, she was just an oddball lady lab rat in a field dominated by men.

One thing she wanted to do with her new technique was
demonstrate that Muller, the rising star in her field, was doing
exactly what he refused to consider and precisely what his friend
Edgar Altenburg suspected. The massively high doses Muller
was using were not, in fact, causing the 'point mutations' he
claimed, meaning the smallest mutations that could trigger
hereditary changes.[2] Muller had even coined the term, still in
use today, but that wasn't what he was inducing in his fruit flies.
The massive doses he was zapping them with were blowing huge
holes in their chromosomes, just like Altenburg and others had
cautioned. And now McClintock could prove it.[3]

Barbara McClintock   **Lewis Stadler**

Credits:
McClintock: https://en.wikipedia.org/wiki/Barbara_McClintock
Stadler: (https://mospace.umsystem.edu/xmlui/handle/10355/66513 and
https://hdl.handle.net/10355/67195 (used with permission)

She was invited to join the University of Missouri faculty by Lewis J. Stadler, a fellow corn geneticist and a rival of Muller who had been irradiating reproductive cells in corn and barley in an effort to induce inherited mutations.[4] Like Muller, Stadler had reached the same erroneous conclusion about inducing point mutations with ultra-high doses,[5] but Muller published first and got the glory. Stadler had been delayed for three months waiting for his paper to be peer-reviewed, while Muller had skipped such niceties and published without review.[6]

Actually, this may have been Muller's only choice, because his paper had no discussion of methods and material, no discussion of other work in the field, and most remarkably, no data to substantiate his claims. Lacking such fundamental details, a peer review would have almost certainly rejected his paper, but the journal *Science* published it nonetheless.

In short, Muller's paper was not a scientific study, but rather an assertion—something that would not be accepted for publication in any serious scientific journal today. But back in 1927, *Science* was privately owned, not overseen as it is now by the

AAAS (American Association for the Advancement of Science). At that time, the owner and editor of the journal was James McKean Cattel,[7] a noted professor of psychology at Columbia and a friend of fellow faculty member Thomas Hunt Morgan, Muller's Ph.D. advisor. As far as they were concerned, Morgan's intrepid grad student had beaten the genetics world to the holy grail of inducing gene mutations at will. Cattel doubtlessly understood that publishing Muller's landmark paper (however flawed it was) would elevate his journal above any other scientific publication.[8] Muller's breakthrough would enable an entirely new level of genetics research; the man would be an immortal.

But not in the way they thought, because as further investigation revealed Muller's point mutation hypothesis turned out to be wrong. And if Muller didn't know it, Stadler certainly did after McClintock finally persuaded him to let her stain his specimen slides. When Stadler saw she was right, he resolved to rebut Muller's work at the Sixth International Genetics Congress at Cornell University in December 1931, a follow-up to the September 1927 conference in Berlin.[9]

During this same period of time, geneticist George Snell, a postdoctoral student and an admirer of Muller and his work, joined Muller's team at the University of Texas at Austin. Muller had his hands full testing fruit flies, and welcomed the chance to have his work extended to mammals. Using laboratory mice, Snell tried to replicate the same genetic mutations that Muller was finding in his flies. But as well-designed as Snell's study was, he failed to show the hoped-for results, even after zapping his mice with a whopping 150 million times background dose.[10]

Snell never cited Muller's work in his paper, and Muller never cited Snell's work in his. Ethical considerations aside, Muller could afford to ignore Snell's contradictory findings, and Snell likely knew which way the wind was blowing in their line of work. Muller had become a rock star, and his LNT model was the new orthodoxy. It was no surprise that their peers largely ignored Snell's work as well.

A major figure in his own right, Lewis Stadler was slated to give the second of three plenary addresses at the Cornell Conference, with Muller to follow as the main attraction. To the astonishment of everyone in the audience (except for Barbara McClintock), Stadler boldly asserted that Muller had confused an observation with a mechanism.

> To state that an induced variation is a gene mutation is not to explain it, but merely to label it.
> —LEWIS J. STADLER

And Stadler didn't just mention this once in passing; he drove the point home in more than a dozen different and quite explicit ways in the course of his presentation. He explained to the stunned audience that when he applied the McClintock technique to his own slides, he could see that, just like Muller, he too had been causing massive gene deletions rather than point mutations. Unfortunately, ladies and gentlemen, the work on radiation-induced genetic effects would have to start over from scratch. And with that mic drop it was Muller's turn to speak.[11]

As you can imagine, Muller was profoundly shaken, his work having just been upended before an international audience of his peers and acolytes. He somehow got through his address, but a few weeks later, back home in Austin and in the midst of an unpleasant divorce, his students found him in a corn field, dazed and wandering after sleeping off an attempted suicide with barbiturates—ironically, he used the wrong dose.[12]

With the help of supportive colleagues and the Rockefeller Foundation (more on them later), Muller secured research positions in Europe and the Soviet Union. During a year he spent in Germany, the science capital of the world at the time, he worked with other brainiacs to develop his 'single-hit' linear no-threshold model based on something called Target Theory, in which a single, powerful emission of ionizing radiation could cause chromosome damage in reproductive cells sufficient to cause a heritable mutation.[13] Muller also spent time at the University of Edinburgh in Scotland, where he directed his Ph.D. candidate Sachi Prasad Ray-Chaudhuri to conduct what turned out to be a seriously flawed experiment that purported to show a linear, or proportional, response to low doses of ionizing radiation.[14]

Muller returned to the US in 1939, shortly before the outbreak of World War II. Among the tide of refugees fleeing the Nazis was Curt Stern,[15] a fellow geneticist and admirer of Muller ever since Berlin; Stern's mother had even translated and typed Muller's paper for the 1927 conference. Stern landed a job with the top-secret Manhattan Project, conducting genetics research at Rochester University.[16] Like nearly every other scientist working on the Project, Stern had no idea what the big picture was, but he was happy to work in his chosen field. He asked the Project to hire Muller, who was given a teaching position at Amherst College, where he also served as a paid consultant for the war effort, exploring the health effects of ionizing radiation. After the war, their work continued under the auspices of the newly-formed Atomic Energy Commission (AEC).

The criticisms that McClintock and Stadler raised in 1931 had not gone away—Muller's LNT model was not settled science and he knew that Ray-Chaudhuri's 1939 study was flawed.[17] To confirm his single-hit theory, Muller had Curt Stern ask Warren Spencer,[18] an LNT advocate and a geneticist at U Rochester specializing in fruit flies, to run an acute-dose study, something that took Spencer more than a year to conduct.[19]

In experiments of this nature, cohorts (small groups) of flies are taken from a large "control" group of laboratory flies—a normal, freely-breeding population of Drosophila representing healthy fruit flies in the wild. With just four chromosomes, the humble fruit fly is ideal for exploring genetic mutations, and their care and feeding costs a pittance. Since they have a lifespan of only about ten weeks, multiple cohorts and control groups were used in the Spencer experiment, which ran from October 1943 to July 1945.

In his study, Spencer isolated various cohorts from the control group and irradiated them with a particular dose at a particular rate, after which he allowed the cohorts to mate. Their offspring were then examined for mutations to determine the heritable genetic effects, if any, of the dose administered to the parents. The difference between the offspring's mutation rate and the natural, in-the-wild "background" mutation rate of the control group would reveal the biologic effect of the dose. An acute dose, as the name implies, is delivered within seconds or minutes, while a chronic dose is delivered over the duration of the experiment, bit by bit.

By zapping fruit flies with massive, one-time doses ranging from a hefty 25 rads to a whopping 4,000 rads, Spencer's acute-dose study showed a linear response. This supposedly validated Muller's single-hit theory, and by implication the importance of cumulative dose over the dose rate. The idea being that whether the entire dose was delivered all at once, or divided into portions and delivered a bit at a time, the end result would be the same. At least, that's what they were trying to prove.

Unfortunately, there was a fundamental flaw in Spencer's methodology that threw his study into question—some of the acute doses were delivered to different cohorts at different rates. One cohort, for example, received 150 rads at the rate of 10 rads per minute, while another batch of flies received 50 rads at the more rapid dose-rate of 15 to 22 rads per minute. This would have been acceptable if the results were reported separately, but Spencer and Stern's 1946 paper on the experiment combined the results, obscuring any distinction between the two cohorts.

Spencer's Table 2[20] is reproduced in Figure 3 below. Let's focus on columns 1, 2, and 6—Dosage (the total dose delivered), Weekly Lots (cohorts), and r/min (rads per minute). In the fourth row down, a dozen cohorts (12 "weekly lots" of flies) were given 150 rads at 10 rads per minute. The 31 cohorts in the next row were given the same dose about twice as fast, at a rate of 15–22 rads per minute.

FIGURE 3: Table 2 of Warren Spencer's Acute-Dose Study

DOSAGE	WEEKLY LOTS	X.V.	M.A.	T.S.D. IN INCHES	r/min
Control	70				
25-r	43	90	3.0	18.5	15-22
50-r	11	90	3.2		10
50-r	45	90	3.0	18.5	15-22
150-r	12	90	3.2		10
150-r	31	90	2.0	18.5	15-22
500-r	14	90	3.0	18.5	15-22
1000-r	5	90	3.2		10
1000-r	20	90	3.0	7.25	91-100
2000-r	2	90	3.0	7.25	91-100
3000-r	2	90	3.0	7.25	96
4000-r	1	90	3.0	7.25	96

Source: http://hps.org/hpspublications/historylnt/episodeguide.html (Episode 9 @ 3:40)

Again, this in itself would have been fine, except that Spencer combined the results from these two sets of cohorts. He did the same with rows seven and eight. Both, for example, were zapped with 1,000 rads, but row seven got zapped at 10 rads per minute while row eight got zapped at 91–100 rads per minute.

With the high doses Spencer administered, it's no surprise that his fruit flies showed a linear, or proportional, response, and an apparently cumulative response as well—the more they got zapped, the more mutations their offspring had. That's what happens at high doses of just about anything, something doctors have known since the Middle Ages. As Paracelsus, the father of pharmacology, explained way back in 1538:

> All things are poison, for there is nothing without poisonous qualities. It is only the dose which makes a thing poison.

Even a perfectly benign and vital substance like water will kill you, if you drink enough of it in one sitting. The same concept applies to high doses versus low doses of radiation, and for much the same reason—the body can only process so much at one time.

For all his education and talent in the field of fruit-fly genetics, Warren Spencer's experiment was improperly designed and executed. And he should have known better—anyone who didn't sleep through science class can tell you that combining cohorts or dose-rates will confound the results of any experiment. Aside from that singular mistake, his study was further compromised by several other methodological flaws.[21] But as compromised as it was, the Spencer study wouldn't be the last mistake on the road to LNT orthodoxy—not by a long shot.

CHAPTER THREE
Ernst Caspari (1945–1948)

IN THE SUMMER OF 1945, as Warren Spencer was wrapping up his acute-dose study, Curt Stern asked their fellow researcher Dr. Ernst Caspari to re-do the Edinburgh chronic low-dose study.[1] Since the early thirties, Muller had been rightly criticized by his fellow geneticists for relying on Ray-Chaudhuri's work. His grad student's efforts had been rife with procedural flaws and improper controls, including basics like failing to ensure the same diet and environmental temperature for both the control group and the various cohorts.

More importantly, there is no mention in the Ray-Chaudhuri paper of any lead shielding to protect the control group incubator from the doses he administered to the cohorts in the test incubator nearby. Lead shielding was standard equipment in the studies that Muller conducted and oversaw, but there is no indication that Ray-Chaudhuri used it, or that Muller seemed to notice. If there was no shielding, the control group could have inadvertently received 24 r (rads), rendering any results meaningless.[2]

Further complicating matters, Ray-Chaudhuri had to conduct the experiment by himself, his only guidance coming from his correspondence with Muller, who was off delivering lectures and attending conferences. The doctoral candidate was in over his head; he admitted as much in his letters to Muller and the result was plain to see in his paper.[3]

For years afterward, Muller had wanted the study done over again to settle the issue once and for all—his twin claims of 'no threshold' and 'cumulative dose' hung in the balance. The 1945 Spencer study had successfully addressed the high-dose range (or so they thought), but the Edinburgh study would have to be re-done to substantiate Muller's low-dose claims. He hoped these two studies could be stitched together to provide the sweeping proof he needed to show that proportional results will inevitably occur, regardless of the size of the dose or the delivery rate. In simple terms—the more radiation, the more bad stuff happens.

Unlike Muller's struggling grad student in pre-war Edinburgh, Ernst Caspari was an experienced Ph.D. entomologist at the University of Rochester, using the latest precision equipment in one of the best labs in the world. And all of it was secretly funded by the bottomless bank account of the Manhattan Project. Spencer's study was first, followed by Caspari's. But after a year of exacting lab work, Caspari was facing an awkward result that would put him at odds with Spencer, their supervisor Curt Stern, and especially Dr. Muller, who was now a candidate for the Nobel Prize.

As Caspari compiled his data in August of 1946 (the first anniversary of the A-bomb attacks in Japan), it became clear that his work had produced no evidence to support Muller's linear no-threshold model in the low-dose range. To the contrary, Caspari's data showed evidence of a substantial threshold, below which there were no observable excess mutations in the offspring.[4]

He had administered a total of 50 rads or 500 mSv (see Nerd Note below) at a continuous rate of about 166 mSv per week for three consecutive weeks to cohorts of adult fruit flies, roughly one-third of their lifespan. This was equivalent to a chronic (ongoing) dose-rate of about 8,600 mSv per year. That's more than 2,800 times the average US background dose of 3 mSv, and well over 100 times the 70 mSv background dose found in some parts of Kerala, a heavily-populated state in southwest India.[5]

> Nerd Note: There are several systems of radiation measurement, so it's easy to get lost in the numbers and jargon. Today, the rad has been largely replaced by the Gray, where 0.01 Gray, or 10 milliGrays (10 mGy), equals one rad. Both refer to the energy in a received dose of radiation. You will also come across terms like rem and mrem (millirem), with one rem corresponding to one rad for biologic purposes.
>
> In the low-dose range (at 100 milliSieverts per year or less), Grays are considered equivalent to Sieverts, which measure the *biologic effect* of a dose. At higher doses, the biologic effect can be "weighted" with a higher Sievert value depending on the type of radiation, whereas Grays strictly measure the received dose, with no adjustment for its biological effect.
>
> A 25-rad dose in Spencer's experiment was equal to 250 milliGrays, 250 mSv, or 0.25 Sieverts. A 4,000-rad dose was equal to 40,000 milliGrays, 40,000 mSv, or 40 Sv.
>
> Seven Sieverts (7 Sv) is the "LD50" dose for humans, meaning a lethal dose of ionizing radiation for 50% of healthy adults. A 4,000-r dose is about six times larger than that. (There will not be a quiz.)

Unlike the Spencer study, which had numerous flaws,[6] Caspari's equipment was properly calibrated, his diet and environmental controls were dialed in, and his control group had a normal rate of background mutation matching that of healthy flies in the wild. But when the dosing was done and he allowed his specimens to mate, their offspring exhibited the same rate of mutation as the non-irradiated control group. This distinct lack of effect showed that there must be some kind of self-repair mechanism at work in reproductive cells, providing a substantial degree of protection from ionizing radiation.

One year before, Spencer had delivered the same 50-rad dose (500 mSv) to his own *Drosophila* cohorts. But rather than spreading it out over three weeks, as Caspari would do, Spencer delivered the entire dose in ten to fifteen minutes, and when he allowed this zapped cohort to mate their offspring had a higher rate of mutation. Caspari's chronic dose of about 166 mSv per week delivered the same total radiation as Spencer's 500 mSv acute dose, but when Caspari dispensed the radiation over a three-week period a threshold was found. Any pharmacist would have instantly understood the outcome: Ingest a bottle of aspirin in one sitting and you'll have problems. Take one a day and you'll be fine.

Viewed together, the Caspari and Spencer studies showed that while Muller's Proportionality Rule applies in the high-dose range, a threshold—and thus a self-repair mechanism—seems to exist at lower doses. And if that were true, then dose-rate would be the determining factor and not the cumulative, or total, dose.

If reproductive cells really did lack the ability to self-repair, as Muller and his supporters believed, then total dose would be the paramount concern. Hits of ionizing radiation would accumulate like unforgivable micro-aggressions, until a safe limit was exceeded and DNA damage occurred. Ed Calabrese calls this the Piggybank Theory: You put a little bit in, and then a little bit more, or maybe a lot more, and when the piggybank is finally filled up, it breaks open and bad things happen.

Fortunately, this doesn't hold true for ionizing radiation, or almost any other potentially hazardous substance. But Muller and Stern were convinced that radiation was somehow uniquely awful in this regard, a sort of sub-atomic Ebola, and now here was Caspari with compelling evidence that they were sorely mistaken. Stern knew this would upset Muller; it meant that LNT was probably BS (bad science), and the public smackdown he got from Stadler at the Cornell convention may not have been entirely out of line.

Decades later, it came to light that Stern had told Robert L Brent, a highly accomplished physicist and researcher of radiation effects, that Stern had telephoned Muller before his Nobel lecture, imploring him to cite Caspari's work and not Ray-Chaudhuri's, cautioning Muller that his hypothesis may be incorrect. But this last-minute effort was dismissed.[7]

It would be hard to overstate the importance of Caspari's discovery. With the ability to self-repair, even a substantial dose of ionizing radiation, damaging or lethal if delivered all at once, could gradually be absorbed by reproductive cells over a period of time without causing permanent damage. There would thus be no heritable radiation-induced mutations, either in the immediate offspring or in subsequent offspring (sometimes mutations skip a generation).

By rights, Caspari's experiment should have been the beginning of the end of LNT.

3.1 The *Drosophila* in the Ointment

When Caspari submitted the results of his year-long study to Stern, he expected the worst. And sure enough, Stern rejected his findings, speculating that Caspari's control group must have had an unusually high background mutation rate.[8] This would have incorrectly shown a threshold by obscuring effects at low doses. But Caspari dug in his heels and dug through the literature, showing Stern that his controls were actually in the middle of the accepted range for background mutation rates in *Drosophila* experiments.

As a *Drosophila* geneticist, Stern should have already known this, and to his credit he relented when Caspari pushed back. But now Stern had a problem with the boss—Muller would not be amused. Caspari's work was supposed to confirm the Edinburgh study, not refute it, but his flies were not cooperating.

They had lived and loved a good third of their lives away bathed in a hefty dose-rate equivalent of 8,600 mSv per year, and their kids had turned out fine.

On November 6th 1946 Stern wrote to Muller, enclosing Caspari's study for review. By this time, Muller had transferred to the biology department at Indiana University (with his salary covered by the Rockefeller Foundation), and was about to sail to Sweden to accept the Nobel Prize. In Muller's reply to Stern, dated November 12th 1946, he expressed concern that Caspari's findings challenged LNT:

> I wonder whether you are having any steps taken to have the question tested again, with variations in technique. It is of such paramount importance and the results seem so diametrically opposed to that which you and the others have obtained, that I think the funds would be forthcoming for a re-test of the matter.[9]

Muller was in a quandary; 'no safe dose' and 'cumulative dose' were the twin pillars of his linear no-threshold model. Since he believed that reproductive cells cannot repair their own radiation damage, he also believed that harm from even the smallest doses would combine to deliver the same amount of DNA damage as a single large equivalent dose. And now, here was a high-quality study conducted by a seasoned scientist, under the supervision of his main man at Rochester, in a top-notch and well-funded lab, showing clear evidence of a substantial safety threshold, right before Muller was about to accept the Nobel Prize for his work that claimed to refute the same idea. And yet, even a phone call from Stern, one of his staunchest supporters, had failed to make Muller rethink his position.

On December 12, one month to the day after he wrote back to Stern, Muller was in Sweden in tie and tails accepting the 1946 Nobel Prize for Physiology or Medicine, and made no mention of Caspari's findings in his lecture to the admiring audience. The most charitable explanation is that Muller didn't have time to read and absorb Caspari's paper before his Nobel speech, but he could have at least mentioned the results.

Instead, Muller withheld the evidence that challenged his work, something most scientists would regard as a serious lapse of ethics, verging on criminality given the circumstances.[10] On top of that, he cited the flawed study by Ray-Chaudhuri—work which he knew was problematic and that Caspari had debunked, and which Stern implored him to reconsider. The Edinburgh study formed the centerpiece of Muller's argument that there is a linear, or proportional, response to ionizing radiation all the way down to zero dose/zero effect. From his Nobel address:

> In our more recent work with Ray-Chaudhuri, these principles have been extended to total doses as low as 400 rads, and rates as low as 0.01 rads per minute with gamma rays. They leave, we believe, no escape from the conclusion that there is no threshold dose.[11]

In light of these bogus findings, Muller called for the adoption of his LNT model for assessing the risks of ionizing radiation. As he explained to the Nobel audience, and the world press in attendance, the standard dose-threshold model, used in the field of medicine since the 1500s, does not apply when it comes to the effects of ionizing radiation on reproductive cells. Declaring this just sixteen months after the atomic bombings of Hiroshima and Nagasaki, he had their undivided attention.

Radiation, Muller explained, posed a different kind of health threat than anything humanity had ever encountered; even a single ionizing event (single-hit theory) could cause damage to the DNA in our reproductive cells. And if one hit wasn't enough to do the trick, further hits would accumulate until a piggybank within the cell breaks open and genetic damage occurs. In starkly sober terms, he informed them that ionizing radiation constituted a hazard to the very progress of human evolution itself.

Thankfully, none of this is true. But thanks to Muller, far too many people still believe that it is. Even our own NASEM, the

US government's National Academies of Sciences, Engineering, and Medicine (formerly the NAS), who should know better by now, still adheres to Muller's antiquated LNT model. We'll save them for last.

CHAPTER FOUR
Delta Uphoff and Curt Stern
(1946–1949)

AFTER THE WAR, Stern, Caspari, and Spencer continued their work with fruit flies for the Atomic Energy Commission, which had also hired the eminent Dr. Donald Charles to study radiation effects on a population of 400,000 laboratory mice. Since the AEC was testing nuclear weapons and marching troops through the fallout,[1] the results of a large mammal study would be greatly appreciated. Thus far, most radiation research had been conducted on fruit flies.

Unfortunately, the Charles study went nowhere. The man was apparently an ultra-perfectionist and impossible to please, so his work was never finalized. The AEC was used to dealing with eccentric scientists, but this was ridiculous and they were in a jam—they needed mammal data and they were stuck with insects. The only other large mammalian study being conducted at the time was the Life-Span Study on the atomic bomb survivors in Japan. But this massive investigation was only just starting up, and wouldn't render useful data for at least a decade. (More on the Life-Span Study later.)

A new mouse study had also begun at Oak Ridge National Laboratory in Tennessee, home of the Manhattan Project, conducted by the highly-respected team of Drs. William and Liane Russell.[2] But like the Charles study and the Life-Span Study in

Japan, the "Mouse House" study at Oak Ridge would take years to yield substantive results.[3] Until then, the Atomic Energy Commission had to content themselves with zapping fruit flies and extrapolating the effects to human health, not the ideal situation when the Cold War was heating up and your citizens and soldiers were worried about fallout.

In January 1948, Caspari and Stern published a paper[4] in the journal *Genetics* on Caspari's chronic dose experiment, but without the hassle of peer review—conveniently, Stern was the editor-in-chief.[5] They did, however, first submit a draft to Muller for a sort of peer review from on high, who suggested some edits.

When Ed Calabrese read their earlier, unpublished draft nearly seventy years later, he found that a key statement had been omitted from the Conclusions section of the published version. The unpublished draft read:

> From the practical viewpoint, *the results presented open up the possibility that a tolerance dose* [a threshold] for radiation may be found, as far as the production of mutations is concerned. This cannot be considered as proved, however, unless the other possibilities of explanation suggest[ed] in the present paper have been excluded. Therefore, no conclusion as to the practical application of these results seems to be permissible at the present.[6] [*emphasis added*]

With this statement removed, Muller's name was added to the paper's acknowledgements and the study was published in Stern's journal. The 1948 paper strikes an odd tone, taking pains to caution the reader that the findings should not be trusted, as if Stern and Caspari didn't believe their own results and were advising the reader to do the same. As Calabrese describes it:

> The entire discussion [in the paper] was all centered around, 'Here's our data, and we don't want you to accept it. Our data differs from the Spencer [study] that showed

a linear dose response. We don't know why it differed, but we don't want anybody to take [our data] seriously and use it until we can figure out why it didn't support LNT.[7]

As Muller hoped, Stern managed to secure more AEC money to test Caspari's findings. But by then the war was over and the talent pool at U Rochester had dispersed, seeking new opportunities in the postwar economy. Stern ultimately selected Delta Uphoff, an inexperienced first-year master's student. Under Stern's supervision, Uphoff conducted three major experiments to test Caspari's findings, with Muller at U Indiana acting as their remote consultant.[8]

Nine months into her first study (1947), Uphoff realized that her control group had a 40 percent lower background mutation rate than fruit flies in the wild.[9] This was a serious problem, since a falsely-low mutation rate in a control group would make the irradiated flies in a cohort seem more sensitive to radiation than they actually were.

In their classified 1947 report to the Atomic Energy Commission, Stern and Uphoff explained that because the background mutation rate of their control group was so "unexpectedly low," their results were "not interpretable." Curiously, the flies they used were from the Muller-5 *Drosophila* strain of fruit flies, a colony of laboratory flies given to Stern by Muller himself.[10]

Stern wrote to Muller, hoping to resolve the difference between Uphoff's control group, which had 0.17 percent of background mutations known as sex-linked lethals, and Caspari's control group, which had a background rate for sex-linked lethals of around 0.25 percent. If Uphoff were going to properly test Caspari's findings, it was critical to resolve the discrepancy, since the background mutation rate was the yardstick by which any effects (or lack thereof) are measured in the cohorts.

Muller wrote back that a proper control group "should contain something like 0.28[percent] of [sex-linked] lethals." This of course confirmed Caspari's "controls"—a general term referring to things like diet, air temperature, and ambient sunlight, aside from background mutation rates, which is the most critical control but not the only one. By confirming Caspari's controls (in writing, mind you), Muller tacitly confirmed Caspari's conclusion that there is in fact a safety threshold in the low-dose range.[11]

Stern's renewed questioning of Caspari's controls was a troubling sign. Just one year before, he had rejected Caspari's paper for having poor controls. But Caspari stood his ground and Stern backed down. If Stern had actually remained unconvinced, he would not have written Muller in November 1946 to tell him that Caspari's work was showing a safety threshold in the low-dose range. But that's just what Stern did, and then he even called him about it, acutely aware that Muller was about to accept the Nobel prize for his work claiming that a safety threshold for radiation does not exist.

This tells us that a year later, in 1947, before Stern had even contacted Muller about Uphoff's controls, Stern already knew that her fruit flies had a lower background mutation rate than flies in the wild. And yet here he was, asking Muller to evaluate his grad student's controls. Perhaps Stern was humoring her, but in any event Muller's written reply was that Caspari's controls were correct, and not Uphoff's—an admission that Muller never made public.

Further snarling the Uphoff study was the fact that the experiment contained two variables, an elementary mistake that Stern should not have allowed. In a series of private letters, Muller had instructed him on a "preconditioning" protocol of an initial acute dose of radiation administered to the female flies to age the sperm retained in the female flies (this was the first variable).

At the time, the ladies were being fed a diet that temporarily prevented them from laying eggs. After preconditioning, their diet was adjusted to allow for fertilization before the chronic-dose experiment (the second variable) was conducted on them.

As it turns out, the acute-dose pre-conditioning technique is associated with a 2.5-fold increase in mutation rate.[12] Which means that the preconditioning step increased the mutation rate observed in the chronic-dose experiment.

A year later, in 1948, Muller acknowledged the correctness of Caspari's controls (and thus the results of his chronic-dose experiment) in private correspondence with Robley Evans, a pioneer of nuclear medicine and a future president (1972–79) of the Health Physics Society.[13] In their exchange of letters, which amounted to a written debate, Calabrese explains that "Muller admitted to Evans that the data from fruit fly research were not adequate to make confident estimates of low dose effects," and that the available "human data were even more inadequate."[13]

Muller asked Evans to keep their correspondence private, and Evans accommodated his request. However, Evans subsequently drafted an article and sent it to more than fifty colleagues for peer review, including a copy to Muller, in which Evans supported a threshold model largely based on Caspari's work. The feedback was strongly supportive, which apparently inspired Muller to publicly rebut Evans's work by publishing a series of papers in which Muller strongly criticized the Caspari study, saying among other things that the controls were aberrantly high.[15]

Muller's private admissions to Stern in 1946, and to Evans in 1948, are deeply concerning, especially coming from a Nobel laureate who asserted the exact opposite in his acceptance speech. Even more concerning is the fact that Delta Uphoff was using the Muller-5 strain of *Drosophila* for her experiments, which for some strange reason had an unusually low rate of background mutations. This raises a pair of troubling questions:

Why was the background rate in Muller's namesake strain of laboratory flies so low? And why would he admit as much to Stern, or to anyone else for that matter, even in private correspondence? (As part of his research, Dr. Calabrese purchased thousands of letters from the Muller estate, including Muller's November 1946 smoking-gun letter to Stern.[16])

One plausible, and disturbing, explanation of a low mutation rate in the Muller-5 strain is that a falsely low mutation rate in a control group of flies would make any irradiated cohort seem more sensitive to radiation than they actually were. And this would make Muller's no-threshold hypothesis seem more like a plausible theory than the unsubstantiated assumption it was.

If this were indeed the case, it would overshadow the error of omission in Muller's Nobel speech. Breeding a strain with abnormally low controls, and giving a sample to scientists working for the Atomic Energy Commission to develop national radiation safety standards for our military and civilians, smacks of scientific fraud. And not only did Muller fail to tell his Nobel audience that Ray-Chaudhuri had been contradicted by Caspari, but he admitted an entire month before his speech, in writing, that Caspari's controls were probably correct.

Dr. Hermann J. Muller, with a Vial of Fruit Flies

Credit: Getty

Even worse, Stern's telephone call to Muller, exhorting him to cite Caspari's work in his Nobel speech and to not rely on Ray-Chaudhuri's work, fell on deaf ears.

Again, the most charitable explanation is that Muller didn't read Caspari's paper in the month before his speech. But even if he didn't get around to reading it before he received his prize, Muller already knew the pre-war Edinburgh study by Ray-Chaudhuri was flawed—he wouldn't have asked Stern to have Caspari spend an entire year running it again if it weren't. And yet, Muller cited the Edinburgh study in his Nobel speech to substantiate his claim that there was "no escape from the conclusion that there is no threshold dose." And this was in spite of the fact that Muller had corresponded with Ray-Chaudhuri at length about several quality and control issues in the Edinburgh experiment.

In their December 1947 classified report to the AEC, Uphoff and Stern speculated that since their original control values were improperly low, "the question may be raised whether at this initial stage of the project it may reflect a personal bias of the experimenter."[17] Since Uphoff was the experimenter and Stern was her supervisor, he was essentially passing the buck to his subordinate in their co-written report, one in which she got top billing. The report tried to walk back the damning statement with a follow-up opinion, surmising that "It is unlikely that this is the case," but it still makes for uncomfortable reading.

Uphoff, however, had apparently taken the blame in stride, seeing as how she signed the report and continued working for Stern; in spite of her botched experiments his funding had not been cut. The AEC still needed a guideline on radiation safety for both the military and the public, and they needed it yesterday. Since the Life-Span Study in Japan and the Mouse House experiments at Oak Ridge would need several more years to produce meaningful results, they had to rely on Stern and Uphoff's fruit flies.

In 1948, something odd happened with Uphoff's second attempt to test Caspari's results: Her control group had the same abnormally low mutation rate, ruining this experiment just like the first one. And now another odd detail has come to light. As Ed Calabrese explains in his March 2024 paper:

> The [laboratory] data of the two key chronic studies of Stern / Uphoff *have never been found* to the present time, now missing for more than seventy years. Yet, the radiation community has cited and relied heavily upon the research of Stern and Uphoff as being critical for the acceptance of the LNT model.[18] [*emphasis added*]

Prior to Uphoff launching her third attempt, Stern accepted a position at UC Berkeley, where she joined him as a lab tech and grad student. Her Berkeley study had the proper controls, including 0.25% sex-linked lethals, but now she had a different problem—the offspring of her irradiated cohorts were showing a mutation-rate response to ionizing radiation that was three times *higher* than even Muller's LNT model would predict, and neither Uphoff nor Stern could figure out why. But the AEC was getting antsy for answers, so Stern had to go with what he had.

4.1 Cool Hand Curt

The man wasn't holding any aces. He had the flawed 1939 Ray-Chaudhuri study, which claimed to show a linear, or proportional, response to chronic low doses with no safety threshold. Countering that, he had Caspari's 1946 re-do of the Ray-Chaudhuri study, in which Caspari found a substantial threshold. And, he had Warren Spencer's 1945 single-hit study, with doses ranging from 25 rads to 4,000 rads (250 millisieverts to 40 Sieverts), delivered in a matter of minutes. Lastly, he had Delta Uphoff's three failed postwar attempts to counter Caspari's results.

Refresher Nerd Note: One rad equals 10 mGy (milliGrays) or 10 mSv (milliSieverts). So 25 rads equals 250 milliGrays, 250 mSv, or 0.25 Sieverts. Similarly, 4,000 rads equals 40 Grays, or 40 Sieverts.

The LD50 dose for humans (a lethal dose in 50 percent of cases) is Seven Sieverts (7 Sv, or 7,000 mSv). Therefore, the 40-Sievert dose that Spencer administered to his fruit flies was gargantuan—about six times greater than the LD50 for humans.

As we saw in Chapter 2, one big problem with the 1944–1945 Spencer study was that even though some of the doses were administered to different cohorts at different rates, the results were improperly combined. Four years later, Curt Stern was hoping to extend Spencer's linear findings at high doses into the low-dose fog bank with solid lab data to back him up, but all he had to go on was Delta Uphoff's three failed studies and Prasad Ray-Chaudhuri's problematic pre-war attempt in Edinburgh. That being the case, Stern was forced to rely on extrapolation rather than certainty, as Muller had been doing since 1927. In the world of science this is called a SWAG, or Scientific Wild-Ass Guess.

That may seem like a harsh assessment, but the independent replication of an experiment is the very essence of the Scientific Method, an unbiased verification of objective reality dating back to the time of Francis Bacon.[19] Without independent confirmation, both Stern and Muller knew that any claim of linear response in the low-dose range would either remain a hypothesis or solidify into a belief. To be sure, LNT was an idea well worth exploring— lives could be at stake. But until it was either verified or disproven, Muller's linear no-threshold model would be little more than a scare story for instilling nuclear fear and establishing overwrought safety protocols, both of which continue to this day.

By way of responding to the AEC's request for answers, Uphoff and Stern published a brief but enormously consequen-

FIGURE 4: SWAG

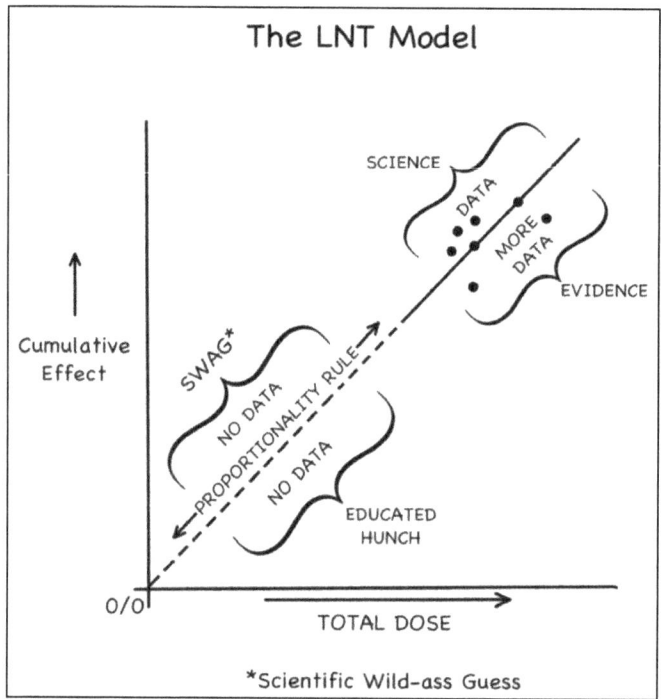

Credit: By the author

tial article in the June 1949 edition of the journal *Science* titled "The Genetic Effects of Low Intensity Irradiation."[20] Their one-page meta-analysis reads as if it were a definitive summation of the research on radiation risk assessment. In it, they rejected the Caspari study for supposedly having falsely-high controls, and accepted the flawed Ray-Chaudhuri study, which claimed to show a linear response at low doses. They also accepted Uphoff's three studies, which had tested Caspari's work and somehow found it lacking. Finally, they accepted the improperly-executed Warren Spencer study, which showed a more-or-less linear response at acute high doses.

They didn't mention in their short report how they had previously admitted to the AEC that Uphoff's first study was uninterpretable, that her second study had the same problem, and that her third study showed abnormally high rates of mutation. They also didn't mention that Hermann Muller himself had confirmed Caspari's controls (and thus Caspari's evidence of a threshold) after Uphoff's first failed attempt to refute the Caspari study, which Uphoff and Stern were now rejecting in favor of Uphoff's work.

Somehow, all three Uphoff studies had now become clearly interpretable, in contradiction to their confidential 1946 report to the AEC. With their public summation in *Science*, Uphoff's three magically-redeemed studies—the first two with data sets that are still missing to this day—supposedly backed their assertion that Caspari's study was bogus and that Ray-Chaudhuri's work at Edinburgh was right on the money. To cap it off, they extrapolated in a straight line from Warren Spencer's evidence of high-dose proportional effects down to Ray-Chaudhuri's claim of low-dose proportional effects, and called it LNT.

The fact that Spencer's study explored chronic doses while the Ray-Chaudhuri study explored acute doses was ignored. And it's worth repeating that both studies, improperly joined at the hip to make LNT seem plausible, had major methodological issues. Undaunted by these vexing details, Uphoff and Stern finished their meta-study with this:

> Viewing all experiments together, it appears that irradiation at low dosages, administered at low intensity, induces mutations in sperm. *There is no threshold below which radiation fails to induce mutations.* A more detailed account of the work will be presented later." [*emphasis added*]

No such detailed account by Uphoff and Stern, together or separately, was ever presented in the journal *Science*, or in any other publication that Calabrese has been able to find. What he did find was that the field of radiation genetics, whose ultimate authority is the National Academy of Sciences (now the National Academies of Sciences, Engineering, and Medicine), continues to cite two key papers in their support of LNT—the flawed 1945 Spencer study and the equally flawed 1949 Uphoff and Stern meta-study. [20]

It's interesting to note that the conclusion of the Uphoff and Stern meta-study refers solely to sperm, rather than both male and female reproductive cells. We'll circle back to this key detail later.

CHAPTER FIVE
Detlev Bronk, Warren Weaver, and Jim Neel (1956)

IN THE 1950S, the atmospheric testing of nuclear weapons was a growing concern, undermining the public's generally favorable view of an atomic-powered future. Radioactive fallout was on everyone's mind, a trending topic that was literally in the air as H-bomb tests became larger and more frequent, while the assurances of the Atomic Energy Commission became less and less reassuring.[1]

FIGURE 6: Mushroom Cloud Comparison Chart

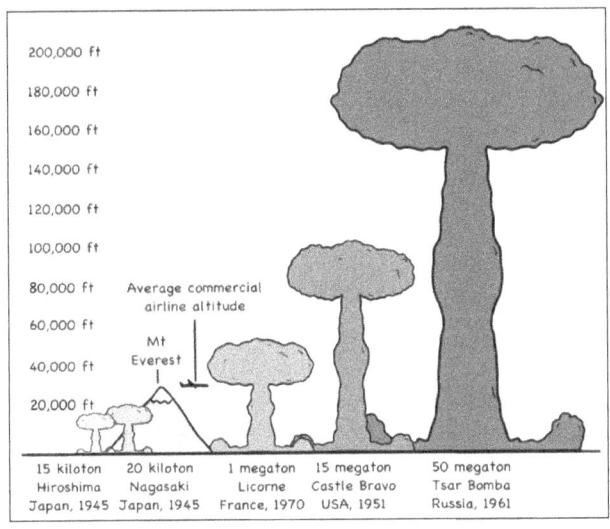

Source: https://9gag.com/gag/awMm3q1

35

With Stadler's untimely death in 1954 Muller's most vocal critic was gone, while Barbara McClintock was being ignored by the male-dominated world of science. As the *Encyclopedia Britannica* explains:

> McClintock's work was ahead of its time and was for many years considered too radical—or was simply ignored—by her fellow scientists. Deeply disappointed with her colleagues, she stopped publishing the results of her work and ceased giving lectures, though she continued doing research. Not until the late 1960s and 1970s, after biologists had determined that the genetic material [that McClintock examined] was DNA, did members of the scientific community begin to verify her early findings. When recognition finally came, McClintock was inundated with awards and honors, most notably the 1983 Nobel Prize for Physiology or Medicine [for her work on jumping genes]. She was the first woman to be the sole winner of this award.[2]

With Stadler and McClintock gone from the limelight, Muller could solidify LNT's position as the principal model for assessing this looming threat to human health. As we saw, one tactic he used was to discredit the Caspari study by claiming the controls were unusually high. This of course contradicted Muller's private correspondence to Stern just a few years before, but the Caspari study had to be discredited to save LNT. This sort of blatant disinformation was a taste of what was to come from Muller and his acolytes in the postwar era of Big Science, spearheaded in the West by the United States.

An interesting detail of this postwar push was that, aside from NASA's moonshot, most of the scientific research and development in this country continued to be funded by the private sector rather than the federal government.[3] The Rockefeller Foundation, for example, had funded some of the wartime research by Ernest O. Lawrence at UC Berkeley, work that led to the atomic bomb. This didn't sit well with everyone at the Foundation, including Rockefeller Foundation president Raymond Fosdick, who was especially troubled by the Foundation's participation.[4]

Awash in petroleum profits, the Rockefeller Foundation had
a long-established and wide-ranging portfolio of philanthropic
endeavors that included funding for research in science and
health. And now, in the midst of an escalating Cold War, the
world was paying close attention to the potential hazards of radiation and fallout; so was the Rockefeller Foundation. Because
as attractive as nuclear power appeared to be, it may be just too
dangerous to consider.

Meanwhile, President Eisenhower sought to reduce global
tensions by promoting the peaceful use of atomic energy. In his
December 8 1953 "Atoms for Peace" speech to the United
Nations, Ike presented an attractive vision of mobilizing scientific experts "to provide abundant electrical energy in the power-
starved areas of the world." [5]

In spite of Ike's inspiring scenario, the Rockefeller Foundation was less sanguine about the prospects of an atomic-powered
future. Abundant energy from nuclear power would not only
increase the world's energy supply, but would also capture considerable market share from fossil-fuel companies, whose stocks
and bonds made up most of the Foundation's substantial endow-
ment. The increased supply of energy would also disrupt the
ability of oil companies to control their output, a mechanism
they used to maintain a favorable balance of supply and demand.
Losing market share and price control at the same time would
cause severe consequences to some large and powerful entities.

The financial threat could be minimized if they could focus
the public's attention on the hazards of radiation and fallout,
and keep it there. And after Hiroshima and Nagasaki, nuclear
fear was an easy button to push. Repeatedly mashing the button
would stymie the development of nuclear power, increasing the
costs for this up-and-coming competitor that seemed to possess
an unbounded potential.

As major funders of scientific and medical research, the Rockefeller Foundation was able to exert an outsized influence on public
perception. And they weren't the only fossil-fuel titans grappling
with this new source of energy. Nuclear power was also a subject

of interest at Shell Oil. One of their premier scientists, Dr. M. King Hubbert, presented a now-famous paper at a March 1956 meeting of the American Petroleum Institute titled "Nuclear Energy and Fossil Fuels" [6] that concluded with a pair of troubling graphs.

Luckily for the Rockefellers and their frenemies in the oil biz, Dr. Detlev Bronk,[7] the president of the Rockefeller Institute for

FIGURE 6: M. King Hubbert's Analysis of Future Energy (1956)

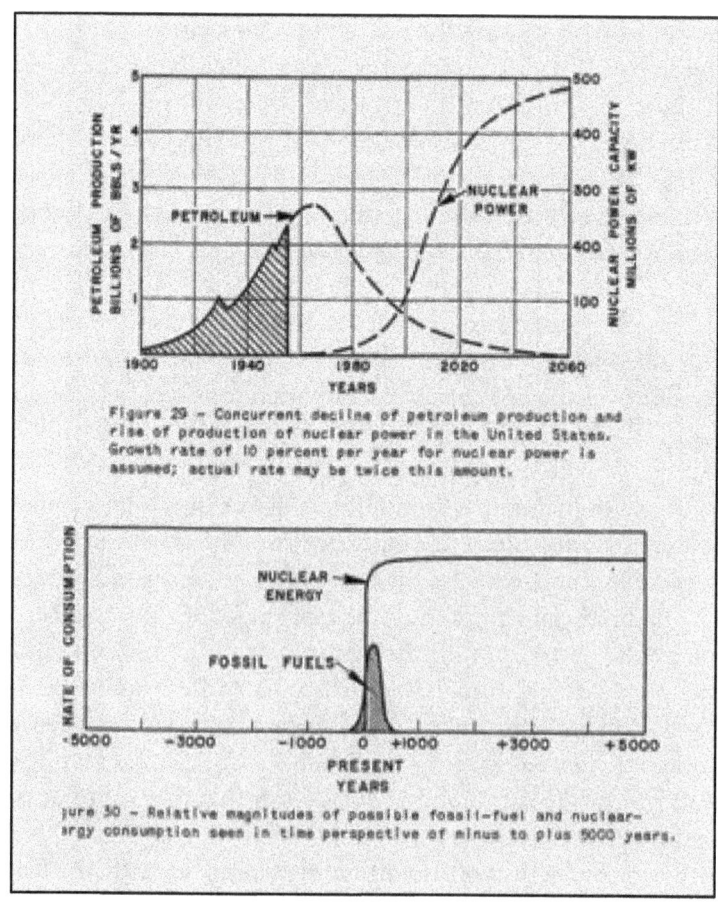

Source: http://stanford.edu/courses/2012/plugis2/docs/1956.pdf

Note: While the exploitation of shale oil and gas has shifted Hubbert's Peak Oil prediction into the future by three or four decades, his science was solid and the implications remain.

Medical Sciences (later Rockefeller University), also happened to be the president of the National Academy of Sciences. In the summer of 1954, a few months after Eisenhower's UN speech, the Rockefeller Foundation's Board of Trustees voted to support a study by the NAS on the biological effects of atomic radiation.

In early 1955, Dean Rusk, a board member of the Foundation who would later become a principal architect of the Vietnam War,[8] wrote to President Eisenhower offering funds from Rockefeller coffers for radiation research:

> Our Trustees wish to contribute to a full exploration of the effects of radiation on living organisms, *with particular attention to the possible danger to the genetic heritage of man himself* . . . Our Trustees wish to act with the utmost seriousness and to make substantial private resources available for the purpose.[8] [*emphasis added*]

With Ike's approval and a generous grant from the Foundation, the National Academy of Sciences formed their first BEAR Committee, with six expert panels to study the Biological Effects of Atomic Radiation. This was the first time in Academy history where they had enough cash to convene a separate genetics panel.[10] In previous studies, they had to put the geneticists on the pathology (medical) panel, where they were routinely out-voted. This time, geneticists would have their own panel and their own panel chair, with an equal voice at the Committee table.

Dr. Bronk appointed Warren Weaver[11] the Rockefeller Foundation's Director for the Natural Sciences, to chair the BEAR-I genetics panel (pronounced "Bear One"). While Weaver was not a geneticist, everyone on the Committee knew who he was—at one time or another he had funded most of them with Foundation money. They were also doubtless aware that the RF had recently bestowed a handsome sum on the University of Indiana for radiation genetics research, which also covered Muller's salary at the college. Additionally, some of them may have known that Muller's stint in Europe, after his debacle at Cornell in 1931, had been financed by the Foundation as well.

Weaver selected a panel of geneticists who all supported LNT, and Muller was the honored member. Among the chosen few was William Russell, who by that time (early 1956) had been running the Mouse House study at Oak Ridge for nearly a decade. Alfred Sturtevant was also on board, a CalTech professor who had worked at Columbia with Thomas Hunt Morgan on *Drosophila* genetics in the 1920s. From his current position at CalTech, Sturtevant was very publicly poking holes in the AEC's position that fallout from nuclear weapons testing was basically harmless.[12]

The second BEAR-I Committee meeting was a doozy. Detlev Bronk asked Warren Weaver, his Rockefeller comrade, to get things rolling. As duly recorded in official NAS transcripts, Weaver explained to the assembled how things worked:

> There may be some very practical results—and this here is the dangerous remark; don't misunderstand me: *We are all just conspirators here together.* I am not talking as an officer of the Rockefeller Foundation, but I will bat my head [against the wall] in the Rockefeller Foundation to try to get a very substantial amount of free support for genetics, if at the end of this thing we have a case for it. *I am not talking about a few thousand dollars, gentlemen. I am talking about a substantial amount of flexible and free support to geneticists.*"[13] [emphasis added]

This surely made them sit up and pay attention. And then genetics panel member Jim Neel cast a shadow over the proceedings by presenting his latest paper,[14] a study that contradicted Muller's hypothesis. Co-authored with William Schull,[15] a geneticist who had worked with Neel in Japan, their paper analyzed the first ten years of real-world radiation health effects data being gathered on nearly 100,000 Japanese blast survivors and their offspring. The ongoing research was being conducted by the Life-Span Study (LSS),[16] a massive US-Japanese joint government project that began two years after the nuclear strikes on Hiroshima and Nagasaki.

As NAS members working with the Academy's Atomic Bomb Casualty Commission (ABCC) in postwar Japan, Neel and Schull had spent nearly a decade engaging directly with the LSS. Starting in 1947, and continuing under the auspices of the Radiation Effects Research Foundation from 1975 until 2012,[17] the sixty-five Life-Span Study monitored the health of nearly 100,000 survivors, along with more than 30,000 non-irradiated Japanese and the descendants of both groups.

With thousands of interviews and on-site observations, LSS staff had determined the particulars of each survivor at the moment of the blast: Their age and health; exactly where they were; what they were doing; what they were wearing; what if anything they were carrying; how far they were from a window or wall; whether there was any other shielding in the radiation path, and so on. By establishing the victims' circumstances in such granular detail, LSS staff could reliably estimate the dose received by each survivor.

In addition to more than 200,000 immediate injuries and deaths, mostly from the shock waves and subsequent firestorms (the cities were largely built of wood and rice paper), the intense radiation from the blasts had consequences for survivors of both sexes, spanning a broad spectrum of age and health. It was a unique, though ghastly, opportunity to study the effects of ionizing radiation, both immediate and long term, on a large cohort of humans and their offspring.

To the consternation of the BEAR-I genetics panel, the Neel and Schull paper showed that a decade after the blast, and contrary to LNT, the offspring of Hiroshima and Nagasaki *hibakusha* (survivors) had not shown any perceptible signs of inherited mutation. From their 1956 paper:

> In summary, then, there emerge from this analysis no *really clear indications that the radiation history of the parents has affected the characteristics of their children* here under

consideration. It should in this connection be pointed out that five of the six findings which give some indications of significance involve the element of maternal exposure, a fact which in view of the possibility of maternal somatic effects suggest the need for particular caution in reaching conclusions.

In order to avoid all possible misunderstandings we hasten to state that under no circumstances can this study be interpreted as indicating that there were no genetic consequences of the atomic bombings. The interpretation is simply that conclusive effects could not be demonstrated.[18] [emphasis added]

In other words, while they didn't find any genetic consequences passed on from parents to offspring, that didn't mean there were none to be found. It was a diplomatic way of not refuting Muller's LNT model, even though nine years of real-world data on a large cohort of humans seemed to be doing exactly that.

But as carefully worded as the paper was, its very existence posed a problem for Muller and his acolytes on the BEAR-I Committee at large, and on the genetics panel in particular. His linear no-threshold model was based on the assumed inevitability of genetic mutations from even the lowest doses of ionizing radiation. And yet, among the tens of thousands of people in Japan who survived the blasts and later had children, there was no discernible evidence of excess genetic defects passed along to their offspring. The unavoidable implication was that the LNT model was probably in error, and that a threshold for radiation probably exists—the kids were all right.

Though the Neel-Schull paper took great care to not reject LNT outright, or even call it into question, their "field work" was scathingly rejected by Muller:

> We should beware of reliance on illusionary conclusions from human data, such as the Hiroshima-Nagasaki data, especially when they seem to be negative [i.e., contrary to LNT].[19]

Everyone got the message, loud and clear. Then it was Tracy Sonneborn's turn to speak, one of Muller's colleagues at U Indiana.[20]

As Calabrese describes it, Sonneborn "read the equivalent of the LNT Apostle's Creed into the record," to the effect that radiation-induced genetic damage is not repairable, not reversible, and that the dose response is linear "down to a single ionization." After Sonnenborn finished, there was no further discussion.[21]

Aside from Jim Neel, whose decade of work in Japan had just been shot down in flames by Muller, no one on the genetics panel had voiced support for a threshold model. As genetics experts, the panel's support of LNT was particularly persuasive. And so it was that in the second meeting of the full BEAR-I Committee, the majority voted for Muller's linear no-threshold model over the pathology panel's counter-proposal to adopt the dose-response model used in pharmacology and toxicology. Both of these long-established disciplines assume that just like any other substance, radiation has a safety threshold below which there is no detectable harm that cannot be routinely self-repaired by a healthy organism.

The pathology panel recommended further research to determine a threshold value for low-dose effects and proceed from there, but they were overruled. Not only the genetics panel, but the entire BEAR-I Committee wasn't about to take any chances with radiation. As Muller had explained in his Nobel speech some ten years before, the future of human evolution was at stake, "the genetic heritage of man himself" as Dean Rusk described it to Eisenhower.

In the decade since his Nobel Prize, Muller's LNT model had become known far beyond the world of evolutionary genetics. Indeed, it was because of Muller that most scientists on the BEAR-I Committee understood just how scary radiation was. To their way of thinking, it seemed awfully unwise to blithely conclude that there were no damaging effects in the low-dose range simply because none had ever been found. Ignoring this potential threat could turn us into a race of mutants.

In the BEAR-I Committee's June 12 1956 pamphlet *Report to the Public*, the NAS affirmed Muller's no-safe-dose model of radiation risk assessment.[22] While the Committee's six panels had their separate technical reports published in *Science*, the

Report to the Public was anonymously written in plain language for general consumption. It's interesting to note that there is no record of anyone on the BEAR-I Committee having signed off on, or even read, the report prior to publication.[23]

The BEAR-I report was launched with a media campaign launched by publisher Arthur Sulzberger's *New York Times*, with a front-page article in the June 12 edition and a free copy of the report sent to every library in the country.[24] Like Detlev Bronk and Warren Weaver, Sulzberger had been on the board of the Rockefeller Foundation for years.[25]

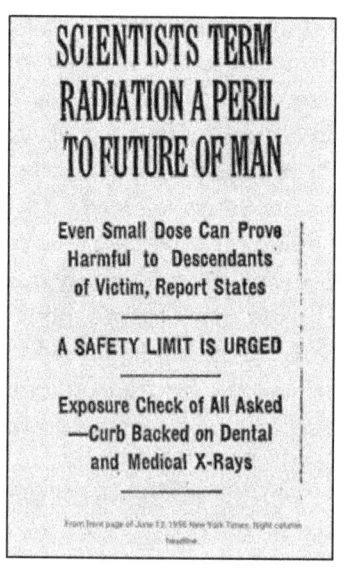

Ever since this pivotal event, a chronic case of nuclear fear has burrowed deep into our collective psyche, taking up residence like an earworm. A cultural bogeyman with staying power, nuclear fear has lingered long after its heyday in the sci-fi fifties. Decades later, popular songs were still raising the specter of genetic damage from radiation, and equating nuclear power with nuclear weapons.[26]

The fix was in.

CHAPTER SIX

Warren Weaver and George Beadle (1956–1957)

IN FEBRUARY 1956, months before the publication of the BEAR-I report, Warren Weaver had a problem—the genetics panel had finished their work in the second meeting, even though they had funding for several more. So Weaver assigned the panel some busy work to give their efforts a veneer of due diligence.[1] It would lend a sense of credibility to their upcoming report, scheduled for publication that summer.

He tasked the twelve geneticists on the panel to estimate the number of birth defects the US populace could expect to suffer from radiation exposure over the next ten generations. They were free to use the methodologies of their own particular specialty, whether they were bacterial geneticists or fruit-fly geneticists or what have you. Weaver's single dictum was that they use LNT as their guiding principle, in which there was no tolerance threshold and that the dose-response was assumed to be linear down to a single dose.

Three panelists declined outright to engage in the exercise, correctly brushing it off as a waste of time—two of them were Jim Neel and, interestingly enough, Tracy Sonneborn. When the remaining nine submitted their homework a month later, panelist James Crow was tasked by Warren Weaver to compile the results. When he did, Dr. Crow saw that his estimate and the other eight were all over the map.

He correctly advised Weaver that this would not enhance the panel's credibility. As Calabrese puts it, "These people who agreed on LNT didn't agree on how to use it."[2] To paper over the problem, Crow removed the three largest outliers and kept the remaining six, which were still scattershot but it was the best he could do. The last six estimates had uncertainties that varied from a 2,000-fold spread down to a confident 10-fold spread; the final report explained it like this:

> These six geneticists concluded, moreover, that the uncertainty in their estimation of the most probable value was about a factor of 10. That is to say, their minimum estimates were about 1/10, and their maximum estimates about 10 times the most probable estimate.[3]

That in itself was accurate, but what the report didn't mention was that while all twelve geneticists on the panel were asked to submit an estimate, three declined and three more had their papers rejected as outliers—a far cry from any sort of consensus or credibility.

With their homework assignments handed in, the genetics panel was itching for something else to do. They even sent a letter

FIGURE 7: Trans-Generational Estimates of the BEAR-I Genetics Panel

Min to Max Range (uncertainty range)	Minimum	Most Probable	Maximum	Author	Basis
1. 2,000-fold	100,000	2,000,000	200,000,000	Beadle	unspecified
2. 2,000-fold	100,000*	4,000,000	200,000,000	Glass	fruit-flies & mice
3. 288-fold	250,000	5,000,000	72,000,000	Crow	humans, mice & fruit-flies
4. 100-fold	600,000	6,000,000	60,000,000	Sturtevant	fruit-flies
5. 100-fold	700,000	7,000,000	70,000,000	Russell	mice
6. 10-fold	2,500,000	10,000,000	25,000,000	Muller	fruit-flies & mice
7.		~~195,000~~		~~Kaufmann~~	~~humans~~
8.		~~50,000~~		~~Wright~~	~~humans~~
9.		~~5,200~~		~~Demerec~~	~~bacteria~~

* On May 21, 1956, Glass wrote to Weaver indicating that his minimum estimate was in error and should be 200,000 rather than 100,000.

Source: http://hps.org/hpspublications/historylnt/episodeguide.html (Episode 12 @ 9:20) to NAS chief Detlev Bronk on August 15th, explaining that some of the things they initially considered taking on were not worth pursuing, and that the BEAR-I public report could probably stand on its own [lightly edited for clarity]:

> We again considered in some detail the question of a fully documented technical report covering the material of the earlier report of the Committee [i.e., the *Report to the Public*]. [But} the more we go into this the more evident it becomes that it would be a big job. It is not at all clear that this effort is justified. For professional geneticists, the literature is already pretty well summarized in several easily available reviews [i.e., the technical papers from each panel]. *For the non-professional, the Academy report seems to serve adequately to summarize the situation.* It is also worthy of note that according to Charles Campbell there have been almost no inquiries about [a more technical] report. This suggests that we are right in *our conclusion that the demand for it is not great.*[4] [emphasis added]

It had been a long, hot summer, and the BEAR-I report rollout hadn't gone as planned. But the dust was finally settling and they didn't want to stir things up again. What happened was that back on June 12th, the same day they released their report, the UK released their own public report on radiation safety. The problem was, the British Medical Research Council's position paper contradicted the US National Academy of Science, and did so by a lot, in plain view of the entire world. The BEAR-I Committee had been in turmoil ever since, and it was clear that Jim Neel was the culprit.

They all knew that Neel had been righteously miffed, ever since the second Committee meeting in early 1956. Put yourself in his shoes: He was an NAS geneticist who had spent the last decade working in postwar Japan on the Life-Span Study with Schull, a fellow NASer. And now Neel was on a dream-team NAS committee, tasked with assessing the biological effects of atomic radiation, and no less than Hermann J. Muller himself,

47

the Godfather of LNT, was pushing back so hard that they wouldn't even look at his damn paper.

Shut down but not silenced, Neel quietly gave his paper to the Brits, who had their own genetics committee compiling their own report on radiation and health, to be published that June in concert with the BEAR-I report.[5] Since the US and UK governments recognized the need for consensus and clear regulations, they planned to inform their people on the same day to bolster confidence among their collective citizenry. As staunch allies in a brewing Cold War, a unified front was of paramount importance.

Instead, Neel and Schull's paper inspired the Brits to stray from LNT orthodoxy. Quite aside from Neel sticking his neck out, this was a bold move for the UK as well. Even before the publication of the BEAR-I report, Muller's risk model had become the world's de facto guideline on radiation safety, in spite of McClintock and Stadler's attempts to correct the bad science. But Stadler had passed away and McClintock had faded from the scene, so the pushback against LNT, at least in the US, was mostly left to Jim Neel.

It was a daunting task, given Muller's prestige and forceful personality, but Neel found allies across the pond. Based on what he and Schull had found in Japan (or rather, what they hadn't found), the British Medical Research Council rejected the no-threshold absolutism of Muller's LNT and recommended a threshold-based pharmacological standard instead. Neel's efforts come through in the BMRC document. For example, paragraph 50 states:

> For the purposes of assessing risk and defining standards of safety, it is necessary to know the nature of the relationship between the dose of radiation and the effect induced. *This relationship may be a simple linear one* in which the incidence of the particular disease increases strictly in proportion to the dose received, *or it may be a curvilinear one* in which, with each successive and equal increment in dose, the incidence increases not by an equal but by a progressively greater amount. *All the evidence suggests that the relation between dosage and radiation effects occurring within a few weeks of expo-*

> sure is of the latter [curvilinear] type, and that the curve shows
> a 'threshold' level, implying that a certain quantity of radiation
> must be exceeded before these particular effects are produced.[6]
> [emphasis added]

While the US concluded that there was no safe dose, their
nuclear allies were recommending a threshold for leukemia at
200 Roentgens (rems) spread over a period of "tens of years."
(The acronym rem stands for "Roentgen equivalent in man," and
"man" in this case refers to healthy adult humans.) From paragraph 255 of the BMRC report:

> We consider, therefore, that an individual could, without feel-
> ing undue concern about developing any of the delayed effects,
> accept a total dose of 200 r [about 2,000 mSv] in his life-time,
> in addition to radiation from the natural background, *provided
> that this* [200 rem] *dose is distributed over tens of years and that
> the maximum weekly exposure, averaged over any period of 13
> consecutive weeks, does not exceed 0.3 r.* We recommend, how-
> ever, that the aim should always be to keep the level of expo-
> sure as low as possible.[7] [emphasis added]

Converting to Sieverts and doing the math: One Roentgen
equals 10 milliSieverts (mSv). A weekly exposure of 0.3 r
(3 mSv) for 13 weeks equals 39 mSv per quarter, or *156 mSv
per year*, totaling 2,000 mSv (200 r) over 12.8 years. This
revealed considerable daylight between the British Medical
Research Council and the US National Academy of Sciences.

Aside from Neel and Schull's work on the Life-Span Study,
the BMRC position was also supported by Caspari's chronic
low-dose study, which suggested that a far heftier 8,600 mSv
per year (166 mSv per week) might even be safe, at least for adult
fruit flies. But Caspari's study had been effectively buried in
1949 by Uphoff and Stern's one-pager in the journal *Science*.
And now here was a lengthy 1956 report from the UK government raising the issue of a threshold all over again.

Even at a comparatively conservative 156 mSv per year, the British recommended threshold was a clear rejection of Muller's no-safe-dose doctrine, which had just been very publicly adopted by the US, with much fanfare in the media. Instead of the Brits sending the same message to their own citizens on the same day, as the two governments had planned, there was now a major disagreement playing out in public between Cold War allies, both of whom were testing nuclear weapons and dusting the world with fallout.

In August 1956, barely two months after this awkward development, Jim Neel gave a convincing presentation contradicting LNT to the First International Congress of Human Genetics in Copenhagen.[8] Muller was in attendance, and tried to prevent Neel from speaking, only backing down when the British geneticists threatened to walk out. The contretemps made it to the *New York Times*, framing the story as a growing conflict between "Mullerians" and "anti-Mullerians," with geneticists around the world choosing sides.[9]

Back in the US, the BEAR-I Committee stood by their report, but behind closed doors the genetics panel was in disarray over the pushback that Neel had stirred up against their well-publicized position. About that same time, George Beadle,[10] head of the biology department at CalTech, was selected to chair the BEAR-I Committee so Warren Weaver could get back to his day job at the Rockefeller Foundation. His work at fellow RF member Detlev Bronk's NAS was done—Muller's LNT model had been officially adopted as the US approach to radiation safety, and free copies of the *Report to the Public* were being widely circulated, establishing LNT as the official word on radiation, even though it disagreed with the Brits to the tune of 156 mSv per year.

With several more Committee meetings still left in the budget, Beadle suggested that the BEAR-I genetics panel explore a question he had posed to his biology department at CalTech the year before. Perhaps they should turn their attention from the issue of radiation and inherited harm, to the issue of radiation and cancer.

Beadle encouraged them to study the health effects that a person might directly acquire from radiation, rather than something they might inherit from an irradiated parent, or something that they, as a radiation recipient, might pass along to their offspring. Specifically, he challenged the genetics panel to explore the possibility that radioactive fallout from the atmospheric testing of hydrogen bombs might be causing cancer.

CHAPTER SEVEN
Ed Lewis (1955–1958)

IN THE 1950S, the California Institute of Technology in Pasadena, California, was a hotbed of environmental activism. Professor Linus Pauling, for example, head of the chemistry department and the 1954 Nobel laureate for Chemistry, was an outspoken advocate for ecology and public health who strongly decried the testing and use of nuclear weapons. An impassioned writer and speaker, Pauling had earned the abiding respect of millions, all around the world.[1]

CalTech was one of several prestigious colleges that received funding from the Rockefeller Foundation, for the express purpose of developing a cadre of scientists skilled in genetics. From the 1958 RF annual report:[2]

> Aid has been given for research in genetics at such leading centers as the California Institute of Technology, Indiana, Texas, Columbia, and Wisconsin . . . [Note that the University of Wisconsin employed Warren Weaver before he worked at the Rockefeller Foundation.]
>
> Grants intended not for any specified set of projects but rather intended to assist in the development of a considerable group of able scientists—*often a whole department or division of biology*—were given to the California Institute of Technology . . . [*emphasis added*]

Radiation was a trending topic, and the atmospheric testing of H-bombs was a growing concern. The blasts were thousands of times larger than the weapons used in Japan, with fallout drifting in the upper atmosphere and dusting the entire planet. In the summer of 1955, George Beadle, chair of the CalTech biology department, challenged his faculty to explore the health effects of fallout on humans.

The one faculty member who showed much interest was Ed Lewis,[3] a young *Drosophila* geneticist in Beadle's department.

Dr. Edward B. Lewis

Source: https://www.facebook.com/6391532964/posts/caltech-geneticist-edward
-lewis-co-winner-of-the-nobel-prize-in-medicine-in-1995/10155633141927965/
Credit: Nolan Patterson

Accepting the challenge, Lewis became particularly interested in the leukemia statistics being gathered by the Life-Span Study in postwar Japan. By this time, the Life-Span Study had been tabulating the health data of blast survivors for nearly ten years, and UNSCEAR (United Nations Scientific Committee on the Effects of Atomic Radiation) was preparing a massive report summarizing the first decade of the ongoing project, to be published in 1958. The LSS, you recall, was the study that Jim Neel worked on for the Atomic Bomb Casualty Commission of the NAS. His summary of Life-Span Study data, co-authored by William Schull, was the paper Muller had sharply rejected in the second BEAR-I genetics panel meeting.

Muller's reasoning was that since field data gathered on humans in real-life situations could be incomplete and misleading, lab testing under controlled conditions was much preferred. To his way of thinking, studying 100,000 fruit flies and their offspring in laboratory conditions could tell us more about irradiated humans than studying 100,000 irradiated humans and their offspring in real-world conditions. Go figure.

George Beadle gave Lewis access to the same body of unpublished Life-Span Study (LSS) data that Neel and Schull reviewed and Muller had rejected. Not only was Beadle the new chair of the BEAR-I Committee, and Ed Lewis's department head at CalTech, he was also an advisor to the Atomic Energy Commission (AEC) and a member of the Atomic Bomb Casualty Commission (ABCC), so Beadle could get his hands on the latest data.[4]

Ed Lewis analyzed the data without collaboration or input from CalTech colleagues, even though he had no expertise in oncology, radiation, epidemiology, or mathematical modeling. Beadle passed around a draft of the Lewis paper to the genetics panel, who offered notes, and shortly thereafter "Leukemia and Ionizing Radiation" by Edward B. Lewis was published on May 17th 1957 in the journal *Science*.[5] Like the BEAR-I *Report to the*

Public the summer before, Lewis's paper landed with a seismic thud that is still being felt today.

The paper incorporated a flawed concept called "somatic mutation hypothesis" first floated by Muller in the 1920s. The gist of the idea was that the low-dose mutation response Muller thought he saw in reproductive cells may also occur in somatic cells—that is, in every cell of the body that isn't a reproductive cell.

Thankfully, the hypothesis was a scientific wild-ass guess that turned out to be wrong. But Muller didn't know this in the 1920s, and thirty years later neither did Lewis. And though Muller was focused on reproductive cells and Lewis was focused on somatic cells, they both assumed that one functioned much like the other.

In some ways they do, but in many ways they don't. Inherited genetic defects and cancer are two very different things, in the same way that reproductive cells and somatic cells are two very different things. For an overview of what science now knows about DNA, somatic cells, reproductive cells, asexual and sexual reproduction (mitosis and meiosis), see the supplement "Sex and the Single Gamete" at the back of this book.[6]

7.1 Mitosis Gone Wild

The bad news in the BEAR-I report was all about DNA damage in *reproductive* cells (gametes), and how this damage could be passed along in the form of inherited mutations. Cancer is an altogether different can of worms, arising from genetic damage in *somatic* cells. The damage is "passed along" in the form of uncontrolled cellular growth in the afflicted person's body, rather than in the person's offspring. And while a predisposition to certain cancers can be inherited, cancer itself is not.

Now that modern science understands the mechanisms of mitosis and meiosis in minute and exquisite detail, it's become clear just how incredibly efficient and resilient our cellular self-repair mechanisms actually are, and how overwrought Lewis's cancer scare actually was. But back when they were doing their work, neither Muller or Lewis could have fully appreciated the complexities of mitosis and meiosis. All they could do was guess.

We now know that the trillions of cells in our bodies have about two meters of chromatin each. These are the spaghetti-like DNA molecules that curl up to form the plump arms and legs of the X and Y chromosomes we're familiar with. If our strands of chromatin were all strung together, they would stretch from the Earth to the Sun and back *more than three hundred times*.[7] Every day of our lives, every healthy cell we have makes about 10,000 repairs on its two-meter allotment of DNA to keep our genes intact.[8]

And that's just routine maintenance; cell division takes more work on top of that, whether it's a somatic cell going through the relatively simple process of asexual division (mitosis), or a reproductive cell going through the two-step process of meiosis (cell division and the preparation for sexual reproduction). Before any kind of cell division actually occurs, sexual or otherwise, the millions of 'base-pairs' in a cell's DNA molecules (the A-T and C-G rungs of the DNA ladder) must first be copied, and then 'proofread'[9] for errors (the animation on this is a must-see[10]). And all this mucking about must be flawless.

Indeed, most cancers are caused by entirely random DNA copying errors that aren't detected and repaired.[11] Despite the occasional hiccup, the replication, proofreading, and self-repair mechanisms found in almost any cell are amazingly rugged and almost foolproof. (We'll discuss the "almost" in the next chapter.)

Damage from something as trivial as a low dose of ionizing radiation is simply taken in stride and repaired on the fly; we are nowhere near as delicate as Muller and Lewis believed. As Ed Calabrese puts it:

> Humans are not victims, but evolutionary survivors. Humans are tough and resilient when damaged, [and will] repair that damage automatically, as these protective processes are built into cells by evolutionary processes. *Humans are not the victims that regulatory agencies would like us to believe.*[12] [emphasis added]

57

Cell repair occurs in the average adult human body more than 3.7 trillion times *per second*,[13] everywhere and all at once, at all hours of the day and night. It's the background hum of life itself, keeping our DNA in shape for both mitosis and meiosis. And while mitosis happens about one trillion times each day in a healthy adult, meiosis is a rare event that plays out deep within the confines of our reproductive organs. This alone affords a high degree of protection, a sort of 'zap me if you can' strategy protected by significant physical safeguards, especially in the female.

Her gamete (reproductive cell) replication and maturation happens in the lower abdomen, shielded by multiple layers of tissue, fluid, and bone from virtually all radiation, other than bombardment from penetrating gamma rays or an internal alpha-particle emitter (see Chapter 2 of *Earth Is a Nuclear Planet*). In the male, gamete production takes place in the testicles, which aren't nearly as well protected. But this greater degree of exposure, with its greater risk of genetic damage, is compensated for in the fusion process (see next chapter).

We now know that healthy living cells can, and do, routinely self-repair damage from low-dose radiation every day of their lives. But even without knowing all the fun details about mitosis and meiosis (see the supplement), cellular self-repair should be obvious to any objective observer, because the simple fact of the matter is that *here we are*, all eight billion of us and counting—the fruit of a family tree that took root some 3.7 billion years ago, in a far more radioactive environment than the one we enjoy today. [15]

Even so, Muller had convinced himself, and the world, that reproductive cells are fragile little things. Extrapolating from there, Lewis reached the same conclusion about somatic cells. Although, to be entirely fair, there is an apparent weak spot in the sexual reproductive process that Muller and his contemporaries were hung up on. So was Lewis, some thirty years later. But the operative word is "apparent." And as it turns out, it's the key to the LNT puzzle.

FIGURE 8: The Family Tree
(We're on the right, next to the slime molds.)

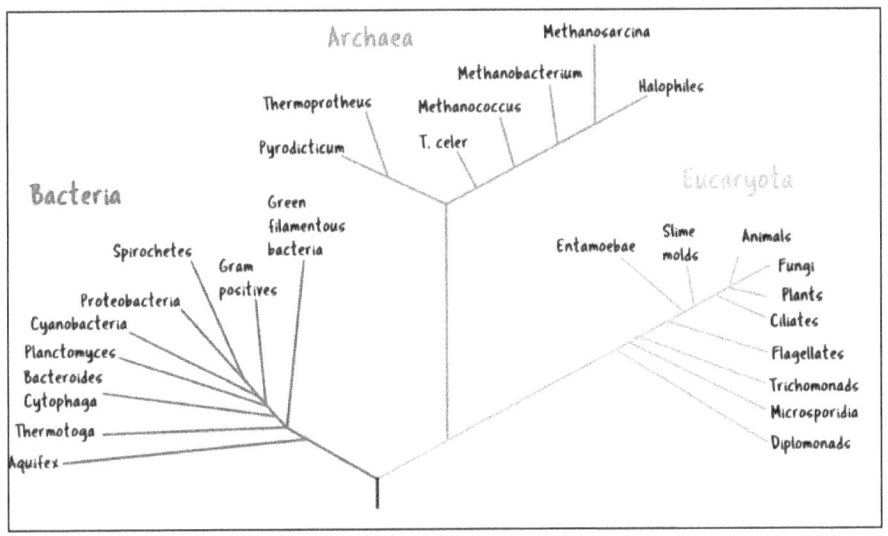

Source:
https://www.researchgate.net/publication/238775969_Microbial_diversity_-_unseen_variety

Knowing what we now know about mitosis, meiosis, and the mechanics of cellular self-repair, this weak spot actually doesn't amount to *bupkis*. But Muller didn't know that, and neither did anyone else at the time. Nor did Lewis, some thirty years later. In the next chapter, we'll find out what it is, why it's inconsequential, and why Muller's mistaken assumption that it's enormously consequential gave birth to the overly-scrupulous safety regime we've been saddled with since the dawn of the Atomic Age.

CHAPTER EIGHT
Metastasizing Muller's Mistake (1957–1958)

Gametes are found in the gonads of sexually-reproducing organisms—the ovaries and testes of animals and the stamens and pistils of flowering plants. The fusion of a mature gamete from a male organism, with a mature gamete of a female organism, results in a unique set of DNA. This is the evolutionary bridge that opened the way to more complex life forms.

IN THE SUMMER OF 1957, Lewis mistakenly crossed that bridge with Muller's no-safe-dose model in tow, and people have been worried ever since about low-dose radiation and cancer. But as it happens, ionizing radiation—especially low-dose ionizing radiation—is one of the least effective ways of triggering cancer. Toxins, pollutants, and the very oxygen we breathe are far more likely to be the culprit.[1] Expanding on his earlier point, Calabrese continues:

> While there are innumerable threats that society faces, the biggest daily personal enemy is ourselves, literally our bodies, which generate *trillions* of adverse genetic events *each day* that must be repaired. It is this *continuous oxidative damage* challenge that we face from our bodies each second, each day, that makes us age, get wrinkly skin, and otherwise get old.

> While people cannot live forever, humans can make far better efforts to extend the health span of our lives. This is not accomplished by avoiding stress but by exposing oneself to a wide range of low-level stresses each day to activate the plethora of adaptive mechanisms that we have been endowed by nature to protect ourselves and optimize health." [2] [*emphasis added*]

Luckily for us, our cells can repair themselves from the onslaught of nearly everything they encounter in life—if they couldn't, you wouldn't be reading this. And it's worth repeating that except in extremely high doses, ionizing radiation is a remarkably weak carcinogen. The fact that this runs counter to what most people believe is a testament to the power of nuclear fear, an all-too-common misperception that's been kept alive for the last seven decades, thanks in no small part to Muller's no-safe-dose BS. Spencer Weart explores the phenomenon in his excellent book *The Rise of Nuclear Fear*. [3]

By the time Ed Lewis was researching his paper in the mid-1950s, reproductive biology was a fairly well-developed body of knowledge. But what was not yet understood was that while female gametes have remarkably strong self-repair mechanisms, their mature male counterparts do not.

From the moment they manifest in a twelve-week old female's embryo, her oocytes (OWE-uh-sites), commonly called eggs, have a full set of self-repair mechanisms to protect themselves from radiation, oxidation, toxins, mutagens, and other stressors at every stage of their lives. Male germ cells, and the gametes that some of them divide into, also have a full set, though not nearly as well-developed as the female's. When immature spermatogonia cells evolve into mature spermatozoa cells to (hopefully) fuse with the ladies, they prepare for their big date by:

- growing a tail so they can go where they need to go
- *shutting down their self-repair mechanisms*. [4]

8.1 Driving Without Insurance

The reproductive process was the particular stage of fruit-fly development that Muller and his contemporaries were focused on, and from what they could see the mature male gamete lacked any semblance of self-repair. Their entirely understandable concern was that if the mature male of the species (of any species, really) could not repair himself, genetic damage could be passed along to any potential offspring. With no self-repair there can be no safe dose, just varying degrees of inheritable mutations.

This apparent weak spot in the male's reproductive process is something that Uphoff and Stern had noted in their one-page meta-study back in 1949:

> Viewing all experiments together, it appears that irradiation at low dosages, administered at low intensity, *induces mutations in sperm*. There is no threshold below which radiation fails to induce mutations." [*emphasis added*]

Notice they didn't mention all gametes, just the male variety.

It was only after the BEAR-I report in 1956, and the Lewis paper in 1957, that this crucial detail of reproductive science was coming into focus in the Mouse House data at Oak Ridge National Lab, which is a story in itself (see Chapter Eleven). In the fertilization process, a tiny male spermatazoa swims upstream and successfully deposits its half-set of chromosomes into a much larger female gamete (egg). These male chromosomes uncurl in their new home, stretching out as long spaghetti-like chromatin molecules, and the female's repair mechanisms get to work.

All this cellular canoodling takes a lot longer than instant noodles—it takes about ninety minutes for the female gamete to diligently inspect and repair any damage found in her newly-acquired set of male chromosomes.[5] And this is why the male's shortcomings don't amount to *bupkis*.

Unless of course the damage is too severe—a lady does have her limits. But even so, you do *not* want the male trying to help her fix his errors. This would lead to mating difficulties, like two jewelers working on the same cuckoo clock. So for a number of very good reasons, Mother Nature has decided that the female, rather than the male, is the one who repairs any damage found, or any that might occur during the fusion event itself. When it comes to something as fundamental as the perpetuation of the species, Ken's just Ken and Barbie's the boss.

Once the repairs are complete, the fertilized egg has a full and proper set of DNA. With a complete set of blueprints, a fertilized reproductive cell is now a somatic cell that can proliferate, all on its own, by asexually dividing into more and more somatic cells to form a new and unique organism, commonly known as a 'baby bump'. And that's the story of the birds and the bees.

A Memorial to Laboratory Mice (Novosibirsk)

Credit: Alamy

Muller irradiated mature *Drosophila* fruit flies with a mega-dose of 130 million times background radiation, and then he mated them to see what he could see in their offspring (he didn't zap them while they were canoodling). He didn't know it at the time, but the wholesale damage he was causing before their fateful rendezvous exceeded the oocyte's capacity to repair both her half and the male's half of their four new fly-baby chromosomes.

Under normal conditions, the female's repair mechanisms are remarkable:

> DNA repair [capability] in sperm is terminated as transcription and translation stops post-spermiogenesis [i.e., when the male gamete's maturation process is complete], so these cells *have no mechanism to repair the damage* [that has] occurred during their transit through the epididymis and post-ejaculation. [However,] *oocytes and early embryos have been shown to repair sperm DNA damage . . .*" [6] [*emphasis added*]

In his effort to be a trailblazing geneticist, Muller created entirely unrealistic conditions in his lab, unlike anything found on Earth. You'd think this would have told him he was barking up the wrong tree, but no. To the contrary, he thought he'd discovered an ironclad principle that applies to all earthly life forms. His friend Edgar Altenburg and other geneticists had their doubts, and argued that he was probably just blasting his specimens with a radiation bazooka. Which, as it turns out, was exactly what he was doing.

Too bad he didn't experiment with tardigrades instead of fruit flies. He could have zapped these microscopic 'water bears' with radiation that was billions of times higher than background, and they would have shrugged it off with a smile. Perhaps Muller would have finally given up, and applied his considerable talent to more useful endeavors. For example, biologists are now exploring how tardigrade proteins could make humans even more resilient to stress, toxins, and radiation.[7]

**The Humble Tardigrade
(Eats Radioactivity for Lunch)**

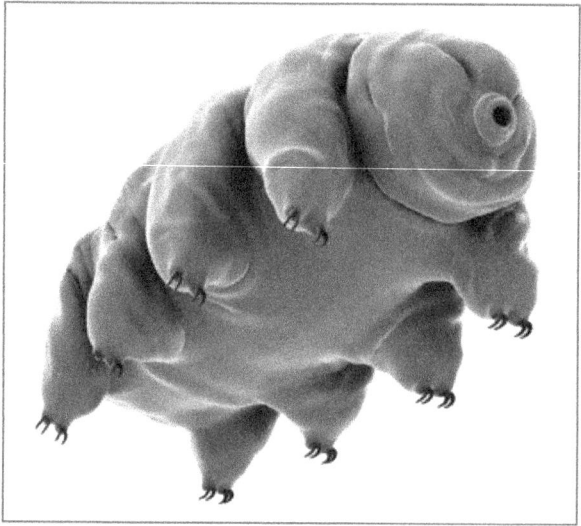

Credit: Alamy

By blasting fruit flies with 130 million times background dose, Muller thought he had discovered that reproductive cells are incapable of radiative self-repair. And at the massive doses he was working with, he was right. The mutations in his *Drosophila* offspring were roughly proportional to the high doses he applied to their parents—doses that would normally be encountered only in a supernova or the core of a nuclear reactor, neither of which anyone had an inkling of back in the 1920s.

Proportional health effects at exorbitantly high doses are no surprise; it's been a well-known phenomenon since Paracelsus, but this was absurd. To make matters worse, Muller improperly extrapolated from the data he found at high doses into unknown territory at low doses, and called it LNT.

Running a straight line through a scattering of high-dose data points, and extrapolating the line down to zero dose / zero effect, is a bold visual statement that effects are directly proportional to the dose received, no matter how small the dose.

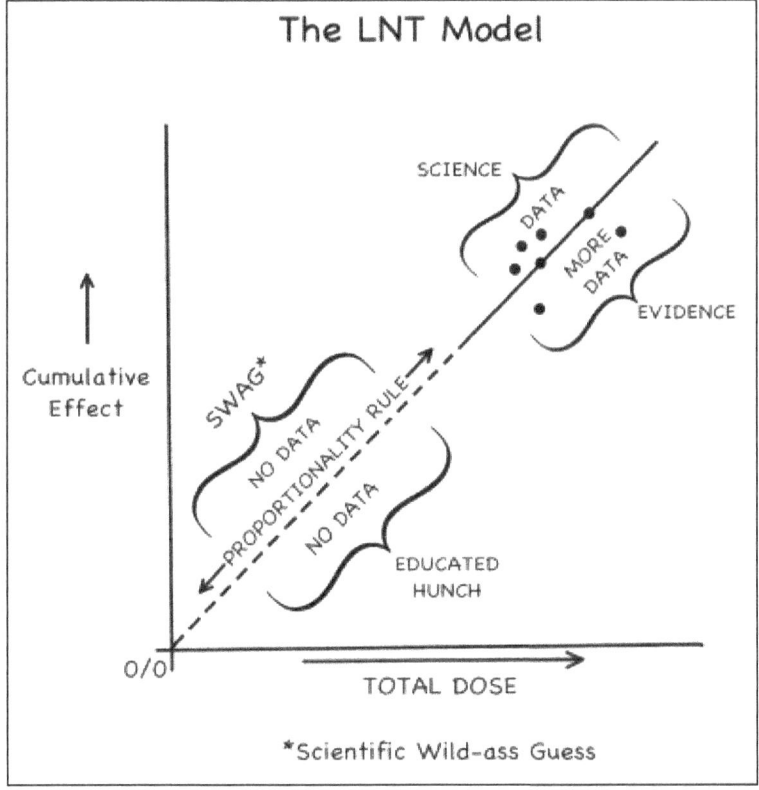

Credit: By the Author

This of course means that there is no threshold, which in turn means that no dose of radiation is safe. But once again, and at the risk of being repetitive—If there was no safe dose of radiation, we literally would not exist. Neither would water bears.

As self-evident as all this becomes when you break it down and think it through, far too many people still adhere to Muller's no-safe-dose doctrine. That's in spite of the patently flawed assumptions behind his discredited idea, and all the advances we've made in biology, nuclear science, and engineering since the 1950s.

Behold the power of nuclear fear.

CHAPTER NINE

From Muller to Lewis:
A Bad Idea Writ Large
(1927–1957)

SOME THIRTY YEARS after Muller's seminal paper, Ed Lewis supersized Muller's principal mistake. To wit: Muller's assumption that all reproductive cells, at every stage of development, are as delicate as mature sperm—repair-challenged, vulnerable, and hoping to score. Taking up George Beadle's challenge, CalTech geneticist Ed Lewis studied the literature, including Muller's foundational work, and came to the erroneous conclusion that low doses of ionizing radiation, from even the faintest trace of fallout, could cause cancer in somatic cells.

In his 1957 paper, Lewis refers to Muller's early musings of a somatic mutation hypothesis, the idea that somatic cells, like reproductive cells, may also lack a self-repair mechanism. If this were the case (thankfully, it's not), then Muller's Proportionality Rule, which evolved into LNT, would be the best way to interpret the data on radiation and cancer.

Lewis had perused the literature on radiation and leukemia of four well-defined cohorts: [1]

- People irradiated for ankylosing spondylitis (spinal arthritis)

- 1,400 infants irradiated for thymus conditions

- Radiologists administering these and other high doses

- The *hibakusha* in Japan.

As the largest of these four groups, the leukemia rates of nearly 100,000 *hibakusha* (blast survivors) would likely yield the most reliable results. The fact that this was the same "field data" rejected by Muller in the BEAR-I meetings didn't seem to dissuade him. Lewis shared George Beadle's interest in exploring this novel threat to public health. It was more than ten years after the nuclear strikes on Japan, and elevated rates of leukemia had been found in the survivors who were closest to ground zero. At the same time, the fallout from nuclear weapons testing was a growing concern.

Leukemia is a cancer of the blood-producing tissue, mostly in the bone marrow where red and white blood cells are made, but it can also manifest in the lymph glands involved in the blood-forming process.[2] The malady has nothing to do with reproductive cells, and most leukemia does not seem to be an inherited predisposition. Infections with certain viruses can cause cancer, and studies suggest that a massive infection (from a bacterium or a vius) can set the stage for cancer later in life.[3]

As for any connection to low-dose radiation or fallout, it turns out that the US Atomic Veterans who witnessed nuclear bomb tests in the 1950s and 1960s, some of whom were marched through ground zero immediately after a blast, subsequently had normal rates of leukemia.[4] But like so much else in the budding field of nuclear science, this was unknown at the time and the concern was entirely appropriate.

Like most other cancers, leukemia seems to result from direct, rather than inherited, effects. There are exceptions, like certain breast cancers, colorectal and prostate cancers, but by and large most cancers come from what we encounter in our own lifetime, rather than something our parents or ancestors encountered in theirs.[5] If Lewis's assumptions about somatic cells were correct, the accumulation of hits from ionizing radiation would be more than just a threat of genetic damage we could bequeath to our descendants, as classical LNT asserted. Lewis expanded the idea to where the tiniest doses of ionizing radiation could also cause DNA damage to our somatic cells, triggering cancer in our own bodies.

He did correctly observe proportional effects in the high-dose range (we'll look at his numbers later). But just like Muller, he improperly extrapolated those effects in a straight line (linear) down to zero dose / zero effect (no threshold), and concluded that every bit of ionizing radiation, no matter how trivial, would contribute to an ever-increasing risk of cancer. Hits both large and small would accumulate over the years, hanging over our lives like a battered piñata, until it finally breaks open and the "Big C" spills out. It was a plausible supposition that fortunately turned out to be wrong.

In the spring of 1957, Lewis gave a draft of his soon-to-be landmark paper to George Beadle, his department head at CalTech and the new chair of the BEAR-I genetics panel. Beadle passed the draft around to other members of the panel, and after some notes from them and some revisions by Lewis, "Leukemia and Ionizing Radiation" was published in the journal *Science*, with a glowing review from Bentley Glass, the editor-in-chief. In the following issue, the magazine published a shout-out from Glass to Lewis:

> In the previous issue E.B. Lewis shows that there is a direct linear relation between the dose of radiation and the occurrence of leukemia, a fatal disease characterized by increases in the numbers of white blood cells. The meaning of such findings is that *any amount of radiation takes its toll on the population and any increase takes a greater toll.*[6] [*emphasis added*]

It's worth mentioning that Dr. Glass, one of the six senior editors at *Science*, was also a member of the BEAR-I genetics panel.

A former grad student of Hermann Muller, Bentley Glass was yet another one of Muller's LNT acolytes.[7]

Meanwhile, Nobel laureate Linus Pauling was making public remarks to the effect that a five-megaton H-bomb test being planned by Britain would probably cause 1,000 cases of leukemia.[8] This got the world's attention, especially from Pauling's fellow scientists, who wanted to know how he arrived at such an alarming number, while Ban-the-Bomb activists stirred with anticipation, awaiting his response. What Pauling was suggesting might be the proof they needed that, contrary to what the Atomic Energy Commission had been telling the public, low-dose fallout from the H-bomb tests was indeed harmful. While it certainly can be harmful, history has shown that most of their concerns were overblown.[9] Unfortunately, their efforts stirred up a generalized fear about all things nuclear, and not just weapons, a social anxiety that has lingered for decades.

Someone had sent Pauling an advance copy of the Lewis paper, and he was troubled by what he read. There was something else he should have been troubled by, but he apparently didn't notice: The paper was a flawed analysis with a pivotal error, and a baseless conclusion that doesn't stand up to scrutiny. (We'll explain why in a bit.)

Before Lewis could even get his paper published, Pauling was creating a media buzz about low-dose radiation and leukemia, a persuasive new talking point in his ongoing campaign against nuclear weapons testing and fallout. Aside from Pauling, fellow luminaries such as Albert Schweitzer, Albert Einstein, and Robert Oppenheimer had also been speaking out against the atmospheric testing of larger and larger bombs by the US, the UK, and the Soviets, and the world was listening.[10] And now this.

The no-safe-dose warning about radiation and genetic defects made headlines the year before, and now in the summer of 1957 the news was even worse, this time about fallout and cancer. The Lewis paper applied the same no-safe-dose assumption to even the slightest trace of fallout, inspiring a concern verging on panic about every speck of dust being lofted on high.

That being said, the atmospheric testing of nuclear weapons was an atrociously bad idea, and not the least because of how far the fallout travels, even if it was (mostly) in the microSievert range and even if there was (almost) no one downwind. Testing nuclear weapons, even underground, is a highly provocative act that does more to elevate tensions than to establish any "balance of terror."

Duck and Cover Drill

Credit: Alamy

Backyard Bomb Shelter

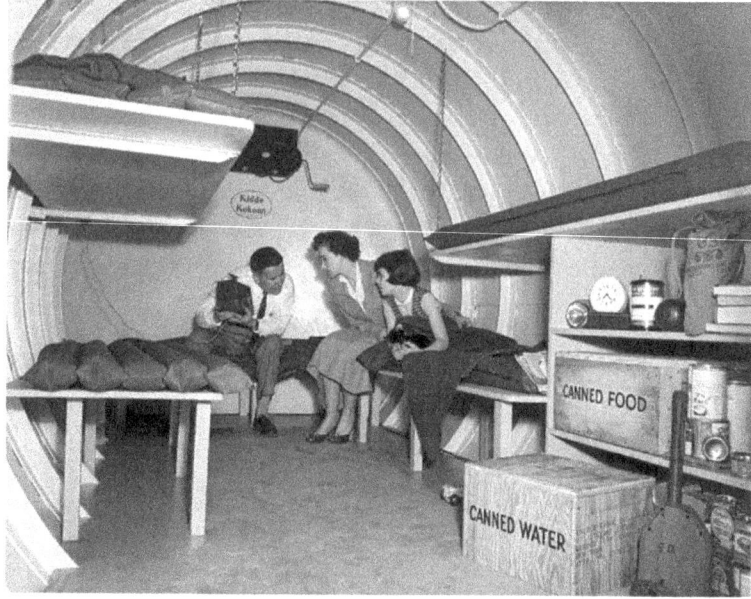

Credit: Alamy

11

Pauling, Lewis, and their supporters knew that a cause-and-effect link between radioactive fallout and cancer would enable them to make an even stronger case against atmospheric testing. Alfred Sturtevant, Pauling's colleague at CalTech and a member of the BEAR-I genetics panel, had already been calling out the AEC for allegedly obscuring the facts, and now he and his fellow scientists had the Lewis paper to press their case.

At the same time, Nelson Rockefeller, soon to be elected governor of New York, was promoting backyard bomb shelters. It was a program he would later pursue as governor, couched in the same grim optimism as the duck-and-cover drills being conducted in grade schools across the country. While the drills didn't cost anything (psychological damage excluded), the billionaire did expect people to pay for their own shelters. This didn't go over well with his fellow New Yorkers.[12]

The health scare promoted by Pauling was that leukemia and

other cancers could be caused by a dispersed cloud of fallout from an H-bomb test on the far side of the world, to say nothing of the blasts being set off in the desert north of Las Vegas. Lewis's paper had convinced Pauling that the concepts of "no safe dose" and "cumulative dose" applied equally to somatic cells, and that damage would accumulate until the recipient came down with cancer.

"Leukemia and Ionizing Radiation" was published in *Science* on May 17th 1957, and Ed Lewis became an overnight media star. The following week, he was featured in the *New York Times* (still helmed by Arthur Sulzberger), and other news outlets snapped up the scary story. Then the May 26th Sunday episode of *Meet the Press* featured Lewis in a contentious exchange with Admiral Lewis Strauss, head of the once-trusted, and now hectored, Atomic Energy Commission (Robert Downey, Jr in the movie *Oppenheimer*). The next day, Monday May 27th, Congress kicked off their hearings on radioactive fallout, with Lewis testifying on June 3rd. A week after that, he was featured in the June 10th edition of *Life* magazine.[13]

The link between fallout and leukemia (and by extension, every other type of cancer) was big news in the summer of 1957. The whole world was paying attention, and not just the science geeks and propeller heads—the LNT scare of the summer before had made the jump from reproductive cells to somatic cells. That is, from mutations in offspring that you may or may not ever have, to mutations in you, courtesy of our friend the atom.

As we'll see in the next chapter, the Lewis paper was BS (bad science). But aside from that pesky detail, the implications were understood by the average person on the street: Radiation is dangerous stuff. Damage from even a single, solitary zoomie colliding with the wrong gene could endanger your life, and it could happen to any cell of your body.

This was LNT on steroids, a summer sci-fi horror sequel in the midst of a very real and expanding Cold War. The Atomic Age wasn't aging well, and the death rays in comic books and sci-fi movies weren't science fiction anymore. Radiation was the new kryptonite, with leukemia and other cancers from low-dose fall-

Zoomies! We're Doomed!

Credit: Wallace Smith / Sara Bancroft

out the Scary New Thing.

And it was all too easy to believe, because the bomb from

whence said fallout came really was the Scary New Thing. The concerns about fallout were, and are, appropriate, but it was all being framed in the most cataclysmic terms. This needlessly freaked out everyone downwind, which at that point in time included virtually everyone on Earth.

Adding to our national anxiety, the Soviets launched Sputnik in October of that same year, a beachball-sized satellite that circled the Earth, beeping a radio signal for three weeks straight until the batteries ran out. Every Western government was listening, while TV stations, radio stations, and ham operators around the world were tuning in to catch the first man-made signal from space.

Once the gee-whiz amazement wore off, the haunting beeps were easy enough to decipher: The bombs were getting more compact, the rockets were getting bigger, and the West was behind the eight ball. It was our first Sputnik Moment.

Stripped of its window dressing, the Space Race that would soon take Americans to the moon was a Cold War scramble to dominate the new high ground. Sputnik had kicked over the chess board, just a few months after we learned that our governments were dusting the world with cancer in their ongoing effort to keep us safe from each other. And it was all playing out in the happy-days background of America's Fabulous Fifties, along with a slew of sci-fi double features down at the drive-in theater.[14]

But now the 1950s were winding down, the Cold War was ramping up, and people were growing increasingly nervous about the prospects of living on the New Frontier.[15] With a "proven" link to cancer and no threshold dose, the obvious solution was to clamp down on anything that emits, or could possibly spread, the slightest hint of radioactivity. The mantra became "lower is better, and zero is always best."

Complications ensued.

CHAPTER TEN
The Atomic Age Hits the LNT Iceberg (1957)

From **MAY 27 THROUGH JUNE 3, 1957,** the Congressional Joint Committee on Atomic Energy convened a special subcommittee on radiation to look into these troubling developments.[1] Ed Lewis was their star witness, offering testimony on the final day of hearings. In the preceding weeks, his paper had generated a blizzard of attention—all this radiation / leukemia business had suddenly become a Very Big Deal. A recent CalTech hagiography about their illustrious alumnus provides the context:

> Lewis was the point person for the linearity hypothesis [LNT] at the Joint Committee on Atomic Energy Congressional Hearings in 1957. Both Linus Pauling, who won the 1954 Nobel Prize for chemistry, and Albert Schweitzer, holder of the 1952 Peace Prize, became active opponents of nuclear testing *in part due to the information Lewis provided in his paper.* With the force of their personalities and the fame generated by their Nobel prizes, they took the statements of many geneticists, including Lewis, to an international audience. *Pauling used Lewis's estimate to predict the number of people that would be killed by leukemia as a result of test detonations,* and informed the public of the magnitude of health hazards. Lewis's work entered the debate over nuclear testing through several

different channels, and it made *a crucial contribution to the scientific and public debates* that led to the Limited Nuclear Test Ban which halted atmospheric nuclear weapons testing in 1963.[2] [*emphasis added*]

All of which is true, and it's great that their efforts led to a ban on atmospheric testing, but their concerns were misplaced about low-dose radiation—of which Lewis's paper had nothing specific to say. This was awkward, because the risk of leukemia and other cancers from low-dose fallout was a central concern of the Joint Committee, not to mention the audience of *Meet the Press*, the readers of *Life* magazine, and the world at large. No one doubted that high doses were harmful. The issue before the Committee was low doses, or rather: Just how low is low enough, Dr. Lewis?

Unfortunately, Lewis could offer no hard data, which is not surprising since he didn't have any. And in any case, he wasn't an authority in the field of radiation safety. Nor was he an expert in public health, toxicology, statistical modeling, medicine, or epidemiology; he was a fruit-fly geneticist.[3] So Lewis offered a seemingly plausible bit of advice that foreshadowed what policy-makers now call the Precautionary Principle.[4] It was the best safety tip he had for anyone driving into the proverbial low-dose fog bank. As he saw it, you had to assume the worst—it was the only sensible thing to do, Senator.

> *The point here, however, is that in the absence of any other information, it seems to me—this is my personal opinion—that the only prudent course is to assume that a straight-line relationship holds here [in the low-dose region] as well as elsewhere in the higher-dose region.*[5]

And with that simple, misinformed hunch from their star witness, the disturbing news about radiation from the National Academy of Sciences the year before had just metastasized, right before the eyes of a Congressional Joint Committee, from a safety hazard for reproductive cells (which was good to know if you were planning a family) into a full-blown cancer risk for every cell of every person's body, from even the smallest dose.

A *Washington Post* headline summed it up nicely: "All Radiation Held Perilous: Nation's Top Geneticists Unanimous in Opinion, Fallout Produced Now Will Shorten Lives in Future, Congress is Told." [6] On June 10th 1957, a week after his appearance before the Joint Committee, *Life* magazine featured a photo of Lewis with this caption:

> WARNING OF DANGER was sounded by Dr. E.B. Lewis of CalTech. In an article in *Science* he proved that there is *a direct relationship between radiation and leukemia.* He predicts a five to ten percent increase in leukemia if strontium-90 levels in humans reaches a figure which the AEC still considers harmless.[7] [*emphasis added*]

By the closing days of 1957, America's Atomic Age confidence was fading fast. Then in early 1958, Linus Pauling's best-selling book *No More War!* characterized the Lewis paper as the "most significant direct information about whether or not small doses of radiation produce cancer in the irradiated human being." [8]

Credit: Scott Medwid

Sorry, Dr. Pauling, but no. What the Lewis paper did was conflate somatic and reproductive cell functions; improperly combine a cohort with a control group (see Chapter 12); extrapolate a high-dose proportional response into a low-dose fog bank; and give the world a chronic case of nuclear fear. *That* was the "most significant direct [mis]information" in the Lewis paper.

And sure enough, Lewis was getting some strong pushback from anti-Mullerians in the genetics field, who didn't find his data at all convincing, mostly because he didn't have any. Even George Beadle, his department head at CalTech, knew that Lewis hadn't proven his claims, and said as much in writing.[9]

But before the karma could catch up to him, his celebrity status landed Lewis a plum appointment to the National Committee on Radiation Protection and Measurements, or NCRP (the M is silent), a Congressionally-chartered private agency founded in 1928. Since the Atomic Energy Commission was being dismantled in light of the ongoing dust-up over H-bomb fallout (thanks in no small part to Dr. Lewis), the NCRP was tasked with recommending radiation protection practices. This wouldn't be too much of a stretch—the Committee was already on record supporting LNT. According to their Handbook 59, published two years before BEAR-I:

> The concept of a tolerance dose involves the assumption that if the dose is lower than a certain value—the threshold value—no injury results. *Since it seems well established that there is no threshold dose for the production of gene mutations by radiation, it follows that strictly speaking there is no such thing as a tolerance dose* when all possible effects of radiation on the individual and future generations are included . . .[10] [*emphasis added*]

But now, in the summer of 1957, there were strong words in the NCRP boardroom about Lewis's flat-footed moment on Capitol Hill. In spite of all the ruckus he stirred up, Lewis could still present no clear and convincing evidence of harmful effects at low doses. Like Muller, he had simply extrapolated the linearity of high-dose effects into the low-dose range and declared his scientific wild-ass guess to be valid.

Speaking of Muller, it must have been awfully nerve-racking for him to hang his reputation on a SWAG. His Nobel speech had given the world a cold-water dip on the subject of radiation, and from that point on he was the Godfather of LNT. But in the four decades between his seminal 1927 paper and his death in 1967, Muller never published any convincing data to back up his model, nor did he publish a single experiment for others to verify his claims. And now here was Lewis, getting a taste of the same pushback.

Arguing out of public view, the NCRP was deadlocked between insisting that Lewis prove there is harm from low doses, and his insistence that they prove there is not. One of his critics at the NCRP was Austin Brues, a former member of the out-voted BEAR-I pathology panel that had recommended the adoption of a threshold dose.[11] But even though Lewis was obliged to admit that he couldn't prove his case,[12] he stuck to his guns and put the onus on them to controvert his findings.

It was a squirrel-cage sort of "logic" reminiscent of Muller's response to Edgar Altenburg in 1927. The flaw in this line of reasoning is that it is often impossible to prove a negative at the subcellular level, since causes and effects can rarely, if ever, be studied in isolation. But since the NCRP was already on board with LNT, a hypothesis that warned of inevitable harm in the absence of absolute safety, it was a short step from there to adopt the precautionary attitude that Lewis offered to Congress.[13] It was a fear-based approach to public policy that boiled down to the *assumption* of adverse effects at low doses of radiation—even though such effects have never been validated after decades of searching, and even though a boatload of empirical evidence suggested otherwise.

Be that as it may, and given the potential dangers involved, Lewis argued that it would be unwise to assume otherwise until everyone was 110% certain, six ways from Sunday. Because you never know ... His monster-under-the-bed approach had nothing to do with nuclear science and everything to do with nuclear fear.

And so it was that by the authority vested in the NCRP, low-dose radiation was pronounced guilty until proven innocent, and nuclear technology would henceforth be in a perpetual state of probation. This was in spite of the fact that just a few months before, in the winter of 1958, a major paper by William and Liane Russell showed a substantial self-repair capacity—and thus a threshold—in highly-irradiated female mice in the Mouse House at Oak Ridge National Lab.[14] (We'll explore this in a bit.)

The NCRP's assumption of harm in the absence of evidence is still in force today. The idea is encapsulated in the catchy acronym ALARA, which recommends that in all things nuclear we should strive to keep radiation down to levels As Low As Reasonably Achievable.[15] This of course raises a question: *Reasonable according to whom?* Establishing an assumption like this about nuclear power (but interestingly enough, not about nuclear medicine) makes it incumbent upon any scientist who disputes LNT to prove a negative. That is, to prove the absence of harm at low doses (i.e., at dose-rates below 100 milliSieverts per year).

Lewis testified again to Congress in 1959 as a high-ranking member of the NCRP, and this time, he had some actual studies to back him up. He cited three papers on leukemia that supported his findings of cancerous effects at low doses, or so he claimed. According to Lewis, the papers showed that high-dose proportional effects can indeed be extrapolated into the low-dose range. From his Congressional testimony:

> *These studies* [exhibits 12, 13, and 14] *and others that have been reported since 1957 have contributed results which are in substantial agreement with the conclusions drawn in testimony presented by the present witness* [i.e., by Lewis himself] *at the 1957 hearing.*[16]

Years later, when Ed Calabrese reviewed the papers, he found that, to the contrary, all three papers actually refute Lewis's claim. Indeed, the studies caution that the proportionality they found at high doses *should not* be extrapolated into the low-dose range.[17] Unfortunately, Congress took the sworn testimony of the witness at his word, without reviewing the evidence—shocking, but true.

CHAPTER ELEVEN
William and Liane Russell, and Paul Selby (1947–1996)

IN 1947, William and Liane Russell initiated a massive, long-term study on radiation effects in mammals. An outgrowth of the Manhattan Project at Oak Ridge, the Mouse House project lasted until 2009.[1] A prominent scientist in his field, William Russell had also served on the 1956 BEAR-I genetics panel, along with Hermann Muller, Jim Neel, Warren Spencer, Alfred Sturtevant, Bentley Glass, George Beadle, and others.

In the Mouse House at ORNL, the Russells had more than a quarter-million laboratory mice at any one time participating in some phase or other of their on-going experiments. Over the course of the sixty-two-year project, the Mouse House used an astonishing total of three million mice, far beyond what any university or private lab could ever hope to wrangle. Like *Drosophila* fruit-fly experiments, the basic idea was to zap mice with various doses of radiation, allow them to mate, and compare their offspring to a control group of non-irradiated specimens representing healthy mice in the wild. It was a simple, straightforward strategy that proved to be surprisingly complex in its execution.

By 1958, the Russells were about ten years into the project. Two years before, William had served on the 1956 BEAR-I genetics panel and supported the adoption of LNT. One year after that, Ed Lewis misinterpreted the LSS data from Japan on the rate of leukemia in blast survivors (more on this in Chapter 12), convincing himself and the world that even low doses of radiation could cause cancer.

Even so, it was becoming clear to the Russells that, contrary to Muller's Proportionality Rule, the oocytes of their irradiated female mice were exhibiting a substantial self-repair threshold. Below a whopping 27,000 times background dose, delivered at 0.09 mSv per minute, the reproductive cells of female mice could adequately engage in self-repair. That's about 50 Sieverts, or about seven LD50 doses, per year. (Here's the math:[2])

Even more remarkable, immature gametes (spermatagonia) in the male mice were also showing a self-repair threshold, though not as high as the female's. It was only when these male gametes matured into spermatozoa to fuse with a mature female gamete, that they seemed to lose all talent for self-repair.

While mature spermatozoa do indeed shut down their self-repair mechanisms, the data that put this curious phenomenon into proper perspective lay buried under a mountain of Mouse House data until the 1990s. And even when the salient facts were finally pieced together, the results weren't officially acknowledged until 1996, a full half-century after Muller received the Nobel Prize for "discovering" that there is no safe dose of radiation.

Despite the near-universal acceptance of Muller's work in the mid-1950s, the subsequent years saw William Russell contradicting him more and more, pushing back against the impact that Muller's ideas were having on public health policy. But by that late date, challenging LNT had become something of a quixotic endeavor. As early as 1963, Muller was on the board of the International Commission on Radiation Protection (ICRP), the international version of the NCRP, and his word on radiation safety, for all intents and purposes, was final.[3]

Muller passed away in 1967, and in 1970 Russell attended the Fourth International Congress of Radiation Research in Evian, France, where he took a stand against the Mullerian position that had permeated the mindset of the global radiation community for nearly four decades.[4] He argued that the no-threshold concept does not apply to mammals, at least to small ones like mice, and that a threshold model should be used instead.

Unfortunately, the Evian proceedings weren't published until 1973. By then, Russell was approaching the mandatory retirement age for federal employees, and would have to quit his work at the Mouse House. In 1977, he was put out to pasture.

11.1 That Seventies Agency

In 1972, the newly-formed EPA (Environmental Protection Agency) was seeking guidance on cancer risk assessment. At the same time, the BEIR-I Committee (pronounced "Beer One") was formed by the National Academy of Sciences to assess the Biological Effects of Ionizing Radiation (hence the name change). Aside from James Crow and William Russell, the first BEIR Committee had a different slate of members than BEAR-I in 1956.[5]

Among their many tasks, the new Committee briefed the fledgling EPA on the findings coming out of the Mouse House. The news, unfortunately, was mixed. While female gametes (eggs) exhibited a robust capacity for self-repair, male gametes (sperm) seemed to have a repair rate of only 70 percent.[6]

What the Russells, the BEIR-I Committee, and the EPA didn't yet know was that when a male hits puberty, about 30% of his sperm cells at any one time are maturing into spermatozoa. And as part of their maturation process, they shut down their self-repair mechanisms to get ready for their Big Date. ("Damaged goods? *Moi*? Shirley, you jest.")

As flawed as the Russells' data set was (see below), their correct observation about the female's capacity for self-repair undermined the linear no-threshold model, and they knew it. William Russell made their findings known to his colleagues on the BEIR-I Committee, and the implications were instantly understood:

If there was a threshold, there must be self-repair. And if there was self-repair, the dose-rate principles of pharmacology and toxicology would apply. And if that were true, then dose-rate was the important factor, and not cumulative dose. And if *that* were true, then LNT would be on very shaky ground.

On the other hand, there was a confounding issue with the male gamete that Russell also brought to their attention. While the female could be irradiated with up to 27,000 times background levels, the mature male seemed to be the delicate one. Taking this into account, the Committee concluded that since safety standards must accommodate the weakest link in the chain, a threshold model for low doses of ionizing radiation could not, in good conscience, be considered at this time.

And there was another thing to consider: The mice in these experiments seemed to be fifteen times more sensitive to radiation than fruit flies.[7] This was a concern, because a greater sensitivity in small mammals may be even more pronounced in large mammals, such as we humans. But years later, when the Russell data was reviewed, it was discovered that this sensitivity in mice only seemed to exist because the Russells excluded something called "cluster mutations" from their calculations (more on this later).[8]

But the BEIR-I Committee didn't know this at the time, and if William Russell knew he didn't say anything. The fact is, this critical flaw in the Mouse House data compilation didn't come to light until the early 1990s, and wasn't formally acknowledged by the Department of Energy until 1996. Working with the erroneous information that was available to them at the time, the 1972 BEIR-I Committee invoked the Precautionary Principle and reaffirmed LNT as the regulatory standard.

Although their approach was understandable from a weakest-link public health policy point of view, their misinformed decision continues to distort the regulatory framework for radiation safety and nuclear power. After taking the Committee's advice in 1972,[9] the EPA finally finalized their radiation standards in 1975[10] (slow computers, we guess). At that time in world history, Western governments were still taking their regulatory cues from the US, so Muller's linear no-threshold model became the global yardstick for radiation risk assessment rather than the UK's curvilinear threshold model.

A few years later, in April 1979, the Three Mile Island nuclear power plant in Pennsylvania suffered a partial meltdown with a minor steam release. No one was harmed by radiation; in fact, the average downwind dose was less than a human chest X-ray.[11] The locals were understandably nervous, but their nuclear fear got cranked up to eleven since "everybody knows" there is no safe dose, and that all doses are cumulative. The EPA even said so.

Nuclear fear spread like a prairie fire, and the embers are still smoldering today—mention Three Mile Island to anyone and note their reaction about what amounted to an unscheduled chest X-ray for the average downwinder. In the years that followed, America cut back on nuclear power construction and built more coal plants instead. Since then, hundreds of thousands of American lives have been shortened by the use of coal.

11.2 Paul Selby

Paul Selby has the distinction of being a former protégé of the Russells, and William Russell's only Ph.D. candidate.[12] After getting his doctorate in 1972, Selby spent three years in Germany, then returned to Oak Ridge National Lab, where he stayed until he retired. His responsibilities at the Mouse House included the migration of their mountain of data onto one of ORNL's first computers.

In the early 1990s, he was asked to migrate the data to the lab's new computers, and as he carried out the project he noticed some irregularities in the control group numbers. With access going back to the first days of the study, Selby quietly sifted through decades of data and came to the disturbing conclusion that the control groups for several Mouse House studies had unnaturally low rates of mutation.

He discovered that cluster mutations, which are common in the natural world, were improperly excluded from control group numbers, even though the raw data showed that clusters were being observed in the control populations as early as 1951.[13] In a colony of healthy mice in the wild, newborn sibling mice can manifest the same mutation. (This is called a cluster—all of a sudden, there's a bunch of mice running around with, say, black ears). But rather than evolving into an enduring feature of the colony, the trait recedes in subsequent generations, like a fashion trend that doesn't catch on.

These random episodes are a natural and well-known feature of mouse genetics, and must be factored into any proper control group's background mutation rate. (This is one reason why a control group must be monitored both before and after an experiment.) By failing to consider the cluster mutations that manifested, and later dissipated, in the control groups used over the course of the study, the Russells had thrown their six-decade project into doubt. The reason why is both simple and enormously consequential:

A falsely low rate of mutation in a control group makes the mutation rate in a test cohort seem falsely high, which makes the self-repair abilities of the cohort seem falsely low. And this makes low-dose radiation seem more dangerous than it actually is.

In 1995, Paul Selby shared his findings with the US Department of Energy, who reviewed the evidence and determined that he had scientific standing. This decision alone was a *very* big deal. The Russells were legends in their own time; their work had been instrumental in setting the stringent rules and regulations on radiation safety in the US, and thus the rest of the world. And now one of their long-time associates had discovered that something was amiss.

The DoE appointed a committee at Oak Ridge to investigate, who called in Selby and the retired couple to testify. After much uncomfortable back-and-forthing, the Russells eventually conceded that Selby had proven their controls were wrong. This of course undermined every statistical result of the entire Mouse House study, since any proper experiment requires a proper control group—the yardstick by which any effects, or lack thereof, are measured.[14]

The committee concluded that the Russells had made a serious mistake and wanted them to correct the record. They instructed Selby and the couple to separately publish their findings for a broad review by a jury of their peers. The Russells published their paper in *Proceedings*, the official NAS journal, in which they conceded that their control group was in error by 120%.[15] Selby meanwhile published two papers in the journal *Genetics*, describing how he found errors that were five to seven times larger than what the Russells conceded.[16]

This was huge. The Mouse House work at Oak Ridge had been the gold standard of laboratory radiation genetics for the last half-century—no one and no institution had ever done anything vaguely comparable. An acknowledged error of 120 percent, in such a comprehensive and foundational long-term study, should have inspired a top-to-bottom review of radiation standards in academia, industry, and government alike, both in the US and around the world.

And then ... *nothing happened.*

Since the 1996 Oak Ridge decision, the Russells' admission of error, and the vindication of Paul Selby, the rules and regulations on low-dose radiation have remained serenely intact. It was as if the Mouse House study was every bit as solid as the 1972 BEIR-I Committee and the 1975 EPA assumed it to be. And this has only served to reaffirm Muller's linear no-threshold model as the guiding principle of radiation risk assessment.[17]

This persistent belief in provably bad science has been used to justify the expensive and time-consuming ratcheting-up of regulations and safety protocols in the nuclear industry, far beyond any good sense. With a doctrine like ALARA (As Low As Reasonably Achievable) and with improved instruments detecting ever more faint traces of radiation, down to the decay of a single atom, there is always a new (and expensive) low to be achieved.[18]

11.3 Ed Calabrese

In 2017, Amherst toxicology professor Ed Calabrese re-worked the numbers from the Mouse House study, using the Russells' conservative 120% correction.[19] While Selby had found much higher percentages of error, Calabrese wanted to see the results by using the Russells' more modest correction instead.

Calabrese found that the oocytes (eggs) of the female mice not only showed a substantial threshold, but the data suggested a strong

FIGURE 9: With 120 percent Russell Correction

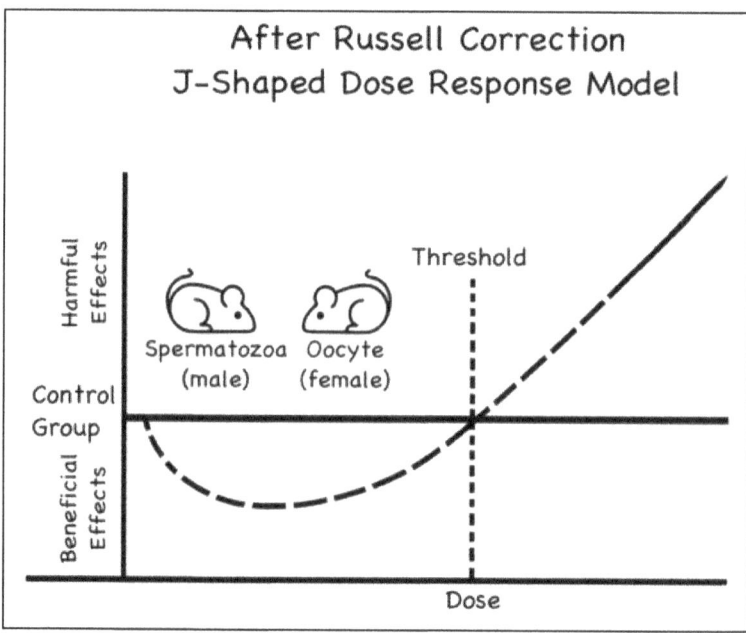

Source: http://hps.org/hpspublications/historylnt/episodeguide.html
(Episode 21 @ 21:02)

protective response as well. In other words, multiple episodes of radiation damage and repair seemed to produce a beneficial, or "hormetic," effect, by stimulating and thus strengthening the adaptive response of the female gametes' self-repair mechanisms. This is not unlike a vaccine exposing the body to antigens—a small, safe dose of a harmful substance that prompts a beneficial response.

Even more surprising, Calabrese also found that the immature sperm (spermatagonia) of the Mouse House males showed a hormetic benefit as well, though not nearly as robust as the female's.[20] In simple terms, the corrected data of the massive sixty-two-year study showed a clear safety threshold for ionizing radiation in both male and female gametes. This was very good news, indeed.

As Calabrese puts it, even the Russells' modest correction had "changed the story entirely."

11.4 There Is No Safe Dose of BS

Unfortunately, none of this came to light until 1996, and by that late date commercial nuclear power in the US was all but moribund, nearly priced out of the market by a morass of overwrought regulations predicated on nuclear fear rather than nuclear science. If the Russells had allowed themselves to work with a data set that included cluster mutations, the world would have known by the 1970s, courtesy of properly-informed BEIR-I Committee, that LNT was bad science and that there was a substantial safety threshold for low-dose radiation in all life forms, including us.

Had this been the case, the newly-minted EPA would have been properly informed by the BEIR-I Committee, and their position on radiation risk assessment would likely have been much different. As a consequence, the public's reaction to 1979 Three Mile Island would likely have been much different from the national panic that ensued. But the paranoia pot was stirred instead, having already been whipped into a froth before the accident by John Gofman and Arthur Tamplin, key personnel at Lawrence Livermore National Laboratory through the 1960s, when they published their 1971 book *Poison Power*.[21]

The book came on the heels of Gofman's testimony in 1970 to the US Senate, in which he raised a false alarm about the supposedly lethal emissions from nuclear power plants in the course of normal operations. The Lawrence Livermore brass were none too pleased with his unsubstantiated claims, and both Gofman and Tamplin soon parted ways with the institution. Even so, their concerns inspired the National Academy of Sciences to launch the first BEIR Committee to study the matter, the upshot of which was the Committee's reaffirmation of LNT.[22] *Poison Power* was reissued shortly after Three Mile Island, and became the New Testament of the anti-nuke crowd.

Had cooler heads prevailed, it would have been clear to one and all that TMI posed no threat to the public. Although it was scary as hell, the reactor suffered a partial meltdown with a small steam release that amounted to a minor industrial accident with no casualties, just some unscheduled chest X-rays. More recently with Fukushima, a similar alternative history could have played out after the accident. The public would have been relieved to know that the highest dose from three full-blown meltdowns at Fukushima—even if no one had evacuated—would have been equal to one whole-body CT scan.[23]

As for Chernobyl, no one will ever build a reactor like that again, with a flammable graphite core and no containment dome. In fact, bringing up Chernobyl in a discussion of nuclear power safety is like bringing up the Ford Pinto in a discussion of automobile safety. The public understands that no car company will ever build a car like that again, and yet every reactor is supposed to be a Chernobyl waiting to happen.

When sufficiently misinformed, we humans can flip out over the most unsubstantiated stuff. And so Muller, Pauling, Lewis, and the Rockefeller Foundation did achieve what they wanted—people around the world clamored to ban the atmospheric testing of nuclear weapons. Which was great, and we applaud their efforts. But with the bad science they employed to make their case, the world they were trying to protect became inordinately fearful of nuclear power as well, with a continued reliance on carbon fuel that has hastened the death of millions.

In *Earth Is a Nuclear Planet*, we conservatively estimated that the coal plants built in the US since TMI (about 45 years ago) have contributed to the premature deaths of over 100,000 Americans. But a study released after we published shows that coal has actually contributed to the premature deaths of at least 460,000 Americans, just in the seven-year span from 1999 to 2007.[24]

Some twenty years after Muller and Lewis made the world afraid of nuclear power, the EPA's overwrought regulations, guided by the BEIR-I Committee, confirmed the public's misplaced concern about every stray wisp of radiation. Even today, nearly a half-century later, the EPA's guidelines on radiation safety are still based on the Precautionary Principle embodied in the 1972 BEIR-I Report, which in turn was based on flawed Mouse House data and Ed Lewis's cancer scare of 1957. And all of it is delicately balanced on Muller's LNT model, falsely validated by his 1946 Nobel Prize and mainstreamed by the BEAR-I report of 1956. In light of this long and nearly continuous stream of disinformation, it's no wonder that some people think nuclear power is a fate worse than global warming.

When the BEIR-I Committee accepted the Russells' erroneous conclusion about mature sperm lacking self-repair, they took a cue from the original BEAR-I report and rejected the Life-Span Study, which by that time had become a twenty-five year compilation of real-world data on nearly 100,000 blast survivors, a non-irradiated cohort of over 30,000 survivors, and the offspring of both groups. This ill-advised decision was in spite of a remarkable anomaly that jumps off the pages of LSS data:

More than twenty-five years after the blasts, Hiroshima and Nagasaki survivors were exhibiting significantly *lower* leukemia rates than Muller, Lewis, or the Russells predicted.[25] (We'll look at this in the next chapter.)

It's interesting to note that during all these years, nuclear medicine has managed to keep out of the fray. Radiologists and radiation therapists have just been quietly going about their business, even in today's anti-nuclear Germany, applying the pharmacological principles of thresholds and dose-rates since they began treating patients with ionizing radiation in 1946—ironically, the same year Muller got the Nobel for his BS on LNT.

After three-quarters of a century of nuclear fear, a glaring cognitive dissonance still exists between the distrust of nuclear power and the acceptance of nuclear medicine. And our national power grid, of all things, is stuck in the middle—an outdated infrastructure on life support, with a long-term prognosis of climate chaos aggravated by excess carbon.

CHAPTER TWELVE
Cuttler and Welsh Revisit Lewis (2015)

IN A DECEMBER 2015 PEER-REVIEWED PAPER,[1] Jerry Cuttler and James Welsh showed that, quite aside from the error of using the science on reproductive cells to evaluate somatic cells, Ed Lewis made an even more egregious mistake, something the world has been stuck with ever since.

In his influential 1957 paper, Lewis had analyzed unpublished Life-Span Study leukemia data on the atomic blast survivors, and made the error of combining two distinct cohorts into a single control group. Using this improperly-blended dataset, Lewis thought he found evidence of a proportional, or linear, relationship between low-dose radiation and leukemia.

It was his flawed analysis, backed by LNT, that laid the foundation for decades of unwarranted fear about low-dose radiation. To fully grasp his mistake, we first need to explore the concept of hormesis and the distinctive 'J-curve' that shows the phenomenon.

Hormesis itself is a familiar idea, particularly in sports medicine, but a controversial concept when applied to low-dose radiation. The term refers to the paradoxical effect in which a small amount of a harmful substance or stressor prompts a beneficial response—basically, a little bit of stress is good for you. Hardly a fringe idea, the hormesis model of "adaptive response" has gained increasing interest in the last few decades.[2]

FIGURE 10: General Hormesis Citations 1999–2014

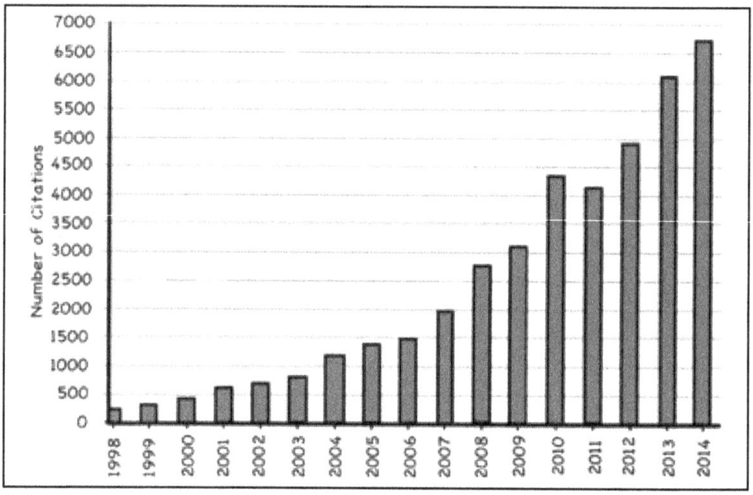

Source:
https://www.toxicology.org/groups/ss/RSESS/doc/RASS-RSESS_Webinar_101415.pdf

Applying the principle of adaptive response to stressors like toxins, oxidation, and low-dose radiation has been hampered by the fact that the pseudo-science of homeopathy borrowed the hormesis idea, and then promptly distorted the original concept beyond all sensibility, with "remedies" diluted to the point where not a single molecule of the therapeutic stressor remains in their potions.[3]

Homeopathic concoctions supposedly work by virtue of the "diluent" (in this case, distilled water) somehow retaining a "memory" of the diluted stressor. To give you an idea of how ridiculously dilute: A typical 30C homeopathic dose is equivalent to administering two billion doses per second to six billion people for four billion years, in order to deliver one molecule of the original stressor to one lucky person.[4]

Most homeopathic remedies contain no measurable, physical amount of any active ingredient that could possibly prompt an adaptive response in the body. At most, they can serve as effective placebos. Hormesis, on the other hand, relies on real and measurable amounts of a physical stressor to "up-regulate" the body's defense mechanisms, enabling a more effective response.

While the two concepts are superficially similar, they have no actual relationship in science or objective reality. Hormesis operates within the bounds of chemistry, physiology, and toxicology, with abundant evidence to support its measurable effects. For example, some beekeepers give themselves pre-season stings, since low doses of the toxin can stimulate the entire system, exercising the body's response ability and thus improving overall health. Homeopathy, on the other hand, dabbles in the realm of magical thinking. Extrapolating from one to the other is not unlike the pretzel logic found in so many other follies.

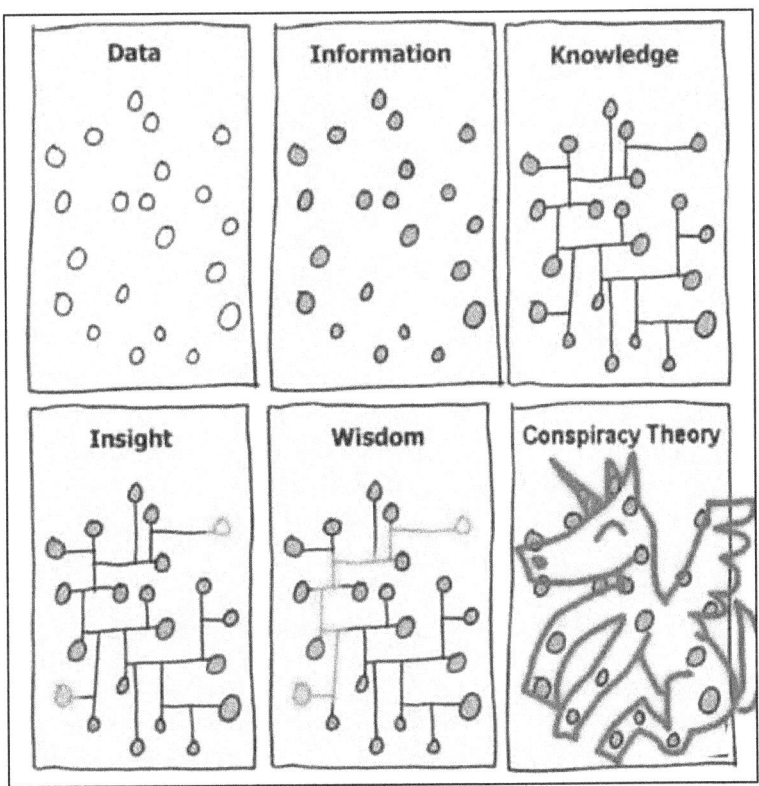

Credit: Gapingvoid.com (used with permission)

Meanwhile, back here in the reality-based world, numerous studies strongly suggest that low-dose radiation has a hormetic effect. The J-curve is a graphical representation of hormesis at work. Note the hormetic model's J-curve in Figure 11 (bottom left), with a dip below the background/control line. In these three graphs, lower is better, so the dip shows that small doses of a stressor can have an opposite, or paradoxical, effect, stimulating the body in beneficial ways. Harm only begins to manifest with higher doses, as the J-curve rises above the control line, or threshold.

FIGURE 11: Radiation Response Models

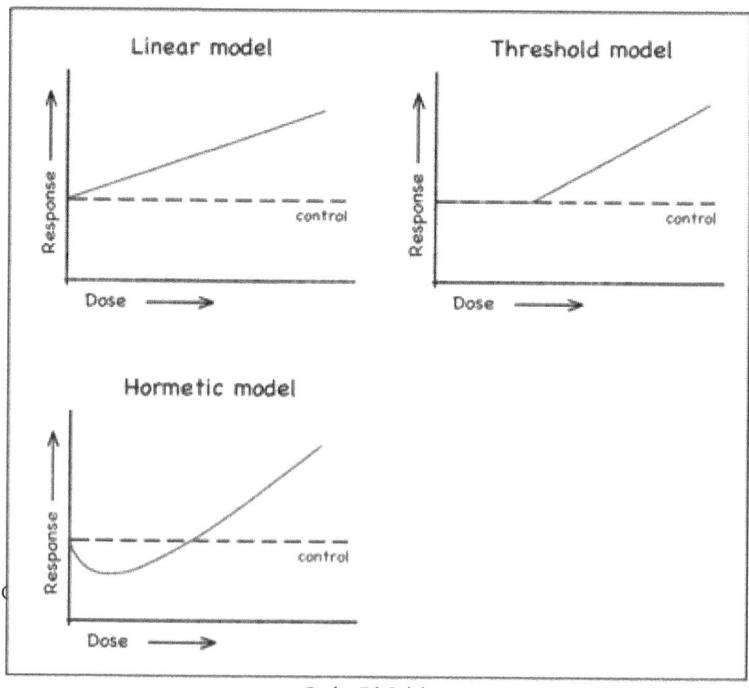

Credit: Ed Calabrese

Cuttler and Welsh showed that when leukemia rates in the blast survivors are properly analyzed by using field data gathered from the original cohorts, there is a distinct hormetic J-curve that reveals a threshold, with an LNT-like linear response only manifesting in the high-dose range.[5] They weren't the first to see a J-curve in the Life-Span Study data, but they were among the first to mainstream this key issue.

In his research on LNT history, Ed Calabrese had come across several studies that showed a J-curve pattern in the low-dose range, papers that had been lost or ignored in the sea of LNT orthodoxy. As the recipient of the 2009 Marie Curie award for his body of work on hormesis,[6] he understood what he was looking at.

The meta-chart in Figure 12 was compiled by Calabrese, showing similar J-curves from nine different studies of the LSS data. Note that the second study on the list is the 1956 BEAR-I pathology panel that recommended a threshold model, but was out-voted by the Muller-dominated genetics panel. Viewed together, the hormetic dip in these nine studies show a clear refutation of LNT, ignored since the 1950s.

FIGURE 12: Nine Studies of the Life-Span Study Data Showing a J-Curve Response in the Low-Dose Range

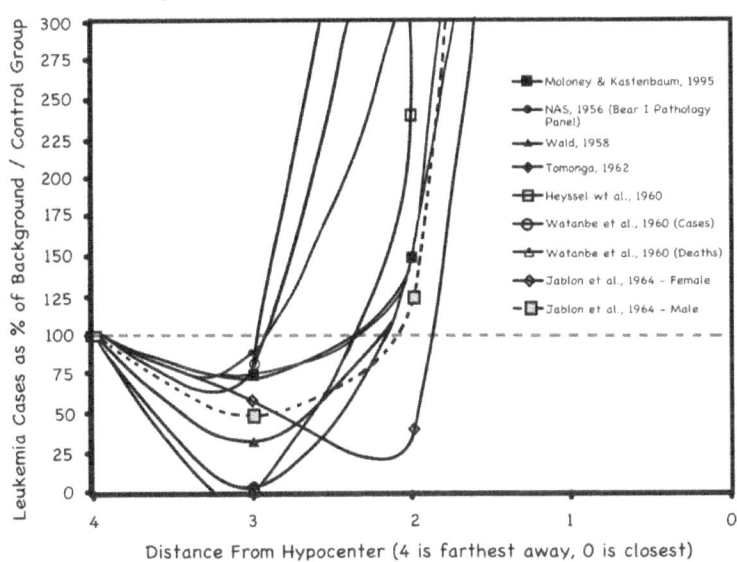

12.1 The Erroneous Zone

We're going to walk you through some sketches, graphs, and tables to show you exactly how Lewis screwed up. It may seem bewildering if you skip ahead and peek at the next several figures, so just stay with us and follow along—we'll keep it simple.

First off, Figure 13 shows how the Life-Span Study organized their examination of the blast sites and survivors. The two bombs, Little Boy and Fat Man, were detonated at an average elevation of 540 meters over Hiroshima and Nagasaki, respectively. The hypocenter (the dot) is the ground-zero point below the aerial blasts. The zones are determined by distance from ground zero.

FIGURE 13: Blast Zone Schematic

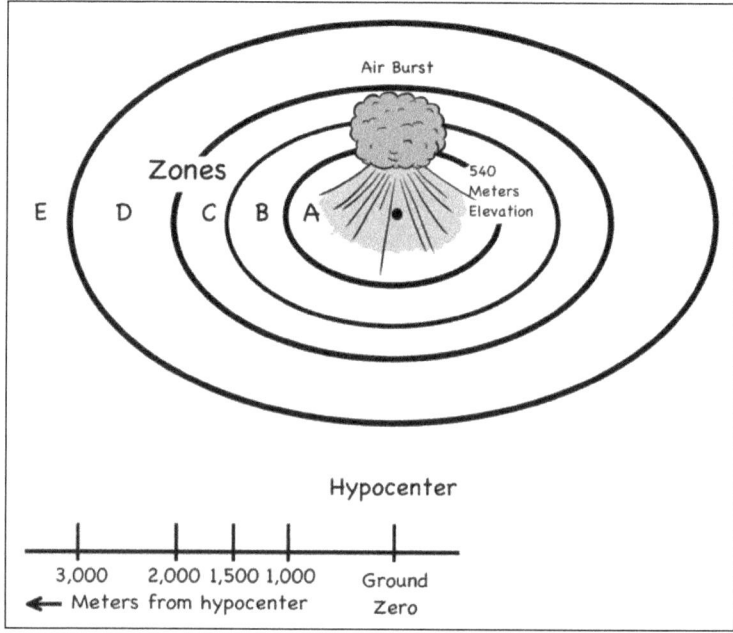

Credit: By the author

The table in Fig. 14 is UNSCEAR's 1958 compilation of the leukemia rates found in the *hibakusha* during the period 1950–1957. This leukemia study was a subset of the Life-Span

Study's broader effort to monitor the blast survivors and their offspring.

FIGURE 14: The Original Five Zones of the LSS Study on Leukemia (UNSCEAR–1958)

TABLE VII. LEUKEMIA INCIDENCE FOR 1950-57 AFTER EXPOSURE AT HIROSHIMA[a]

Zone	Distance from hypocenter (meters)	Dose (rem)	Persons exposed	L (Cases of leukemia)	$\sqrt{L}$	N[b] (total cases per 10^6)	N₄ (Radiation-induced cases per 10^6)	N₄/rem	PL ($N_4/10^6/year/rem$)
A	under 1,000	1,300	1,241	15	3.9	12,087 ± 3,143	11,814	9.1	1.14 × 10⁻⁴
B	1,000-1,499	500	8,810	33	5.7	3,746 ± 647	3,473	6.9	0.86 × 10⁻⁴
C	1,500-1,999	50*	20,113	8	2.8	398 ± 139	125	2.5	0.31 × 10⁻⁴
D	2,000-2,999	2	32,692	3	1.7	92 ± 52	−181	−90	−11 × 10⁻⁴
E	over 3,000	0	32,963	9	3.0	273 ± 91	Control	—	—

* Based on data in reference 13. Prior to 1950 the number of cases may be understated rather seriously.
[b] The standard error is taken as N ($\sqrt{L}$/L).

[c] It has been noted [13, 16] that almost all cases of leukemia in this zone occurred in patients who had severe radiation complaints, indicating that their doses were greater than 50 rem.

Source:
https://www.unscear.org/docs/publications/1958/UNSCEAR_1958_Annex-F-G-H-I.pdf

We'll discuss the circled set of numbers in a bit. But first, let's look at a table from Ed Lewis's 1957 paper (Figure 15 below), and compare it with the UNSCEAR table above. As you can see in the Lewis table, he combined Zones D and E from the UNSCEAR table into an expanded Zone D control group, which we call his Erroneous Zone. This was his principal error, and it is difficult to overstate how consequential it was.

Notice the distances from the hypocenter listed in the first column of the UNSCEAR table: Zone D is 2,000 to 2,999 meters from ground zero, and Zone E is 3,000 meters and beyond. Compare this with Lewis's table below. He describes Zone D as being from 2,000 meters to everything beyond. This improperly includes Zone E, which is actually the correct control group, comprised of people farthest from the blasts who received virtually no radiation at all.

Blending these two zones into one big control group obscured any difference in health effects between these two distinctly different cohorts—the survivors in Zone D who received an average of 20 mSv of ionizing radiation, and those in Zone E who didn't receive any.

FIGURE 15: Ed Lewis's 1957 Table 2 (Note the absence of Zone E)

Table 2. Incidence of leukemia among the combined exposed populations of Hiroshima and Nagasaki by distance from the hypocenter (January 1948–September 1955).

Zone	Distance from hypocenter (m)	Estimated population of exposed survivors (Oct. 1950)	Number of confirmed cases of leukemia	Percentage of leukemia
A	0– 999	1,870	18	0.96
B	1000–1499	13,730	41	0.30
C	1500–1999	23,060	10	0.043
D	2000 and over	156,400	26	0.017

Source: Ed Lewis 1957
https://link.springer.com/chapter/10.1007/978-1-4020-6345-9_21

Figure 16 is Cuttler and Welsh's correction of Lewis's Table 2, in which they restore the original five zones established by the Life-Span Study and reported by UNSCEAR in 1958. This same data was available to Lewis as he was writing his paper, courtesy of his CalTech depart-

FIGURE 16: Cuttler and Welsh's 2015 Correction of Lewis's Table 2

Zone	Distance from hypocentre (m)	Dose (rem or cSv)	Persons exposed	Number of cases of leukemia	Total cases per million
A	0 - 999	1300	1,241	15	12,087
B	1000 - 1499	500	8,810	33	3,746
C	1500 - 1999	50	20,113	8	398
D	2000 - 2999	2	33,692	3	92
E	over 3000	0	32,963	9	273

Table 2: Leukemia incidence for 1950-57 after exposure at Hiroshima (adapted from UNSCEAR-1958, Annex G, Table VII) [8].

Source: Cuttler and Welsh 2015
https://www.researchgate.net/publication/338819152_Leukemia_and_Ionizing_Radiation_Revisited

ment head George Beadle. With such primo access to the latest LSS data, the alteration in Lewis's Table 2 is something of a mystery.

> Nerd Note: One centiSievert (cSv) equals 10 mil-lisieverts (10 mSv) which equals one rem. So the 50 cSv dose noted for Zone C means that an average of 500 mSv was received by this cohort.

We'll explore the circled numbers in a bit (they're the same numbers we circled in the UNSCEAR table). But first, let's take a moment to review:

When Lewis was researching his paper, the Life-Span Study had been an ongoing activity for a little over a decade, gathering data on the health of blast survivors in postwar Japan. The first nine years of the data were examined by Neel and Schull in their 1956 paper, which was dismissed by BEAR-I genetics panel member Hermann Muller as unreliable field work. But after an awkward public disagreement between the US and the Brits about low-dose safety thresholds (or a lack thereof), George Beadle, the new head of the BEAR-I genetics panel, was giving fellow CalTech professor Ed Lewis access to the same data.

As Jim Neel's paper had tactfully explained the year before, the Life-Span Study data contradicted LNT by showing no excess inherited genetic mutations in the offspring of nearly 100,000 blast survivors from zones A through D, and in more than 30,000 survivors from Zone E. But Lewis was focusing on radiation and cancer, rather than radiation and inherited muta-tions, and in the wake of the BEAR-I report's awkward rollout vis-à-vis the British position paper, the BEAR-I genetics panel felt that Lewis was onto something that might exonerate their stance on radiation safety. And indeed he was, but he got there by obscuring a crucial distinction with a serious mistake.

12.2 Hormesis and the J-Curve

Today's leukemia rate in the US is 140 cases per million people,[7] about half the rate of wartime Japan. After years of imperial aggression, first in China, then in Southeast Asia and the Pacific, followed by months of carpet-bombing in the final stages of the war, the Japanese homeland was suffering from malnutrition and disease and less than 20% of them had refrigeration. To make matters worse, the average adult in postwar Japan smoked about four cigarettes a day.[8]

Leukemia is strongly associated with a prior massive viral infection.[9] Infections cause inflammation, and chronic inflammation is associated with cancer. It is also known that DNA repair mechanisms are impaired in malnourished persons.[10] Because of these factors and others, the state of public health can be a significant predisposition for leukemia and must not be ignored. But none of these confounding factors are discussed in Lewis's paper.

The Zone E control group established by the Life-Span Study consisted of nearly 33,000 non-irradiated Japanese from the outlying Hiroshima and Nagasaki regions. These were people who survived the nuclear attacks, many of whom were still alive during the 1950–1957 leukemia portion of the study. As we saw in Figures 14 and 16, the blast survivors from Zone E had a subsequent leukemia rate of 273 cases per million. This was the background, or control, rate of leukemia in the non-irradiated postwar population, and served as the yardstick by which leukemia rates in the other Zones should be compared.

According to Muller's Proportionality Rule (the underlying concept of LNT), the leukemia rate in Zone D, which was closer to the blast than Zone E, should be proportionally higher, and the rates proportionally higher still for zones B, C, and A. But that's not what happened.

The Zone E survivors, representing the broader population, had a leukemia rate of 273 cases per million twelve years after the blasts. But the survivors from Zone D, who were closer to the blasts and received an average of 20 mSv (2 rem) of radiation, had a significantly lower leukemia rate, at just 92 cases per million. That's a reduction of 66 percent.

Moving closer to the hypocenter, the hormetic benefit ceases in about the middle of Zone C and the leukemia rate returns to

273 cases per million. The survivors in this zone received an average of 500 mSv (50 rem) and subsequently had an average leukemia rate of 398 cases per million. As we move even closer to ground zero, the case-rate rises linearly, as LNT predicts.

This telltale dip—from 273 down to 92, then back up through 273 to 398 and beyond—is a classic J-curve response to the relatively mild dose of ionizing radiation in Zone D.

FIGURE 17: The J-Curve Hormetic Model

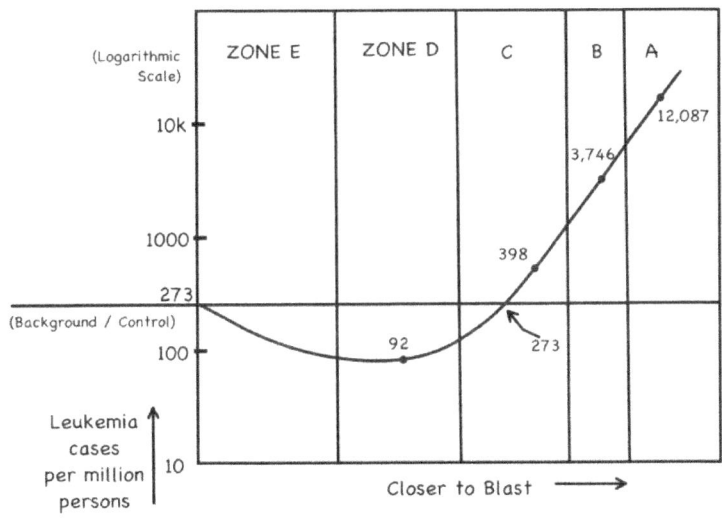

Source: *Ed Calabrese*

12.3 Don't Be Lewis!

The paradoxical effect of a reduced cancer rate associated with a higher (but still low) dose of ionizing radiation is impossible to explain with LNT. Intentionally or not, Lewis obscured critical evidence by grouping Zone D with Zone E to form his Erroneous Zone. It was such a remarkable mistake that Cuttler and Welsh quote his reason for doing so from his 1957 paper:

> Since the majority of the population in Zone D (from 2,000 meters on) was beyond 2,500 meters, the average

dose is under 5 rem and is thus so low that *zone D can be treated as if it were a 'control' zone."* [*emphasis added*]

No, it can't. And this is where Lewis went off the rails. On top of that, the average dose in the outer portion of Zone D was about 1 rem, or 10 mSv, so he got that wrong as well. But his greatest mistake was combining the two zones into an expanded control group.

In 2015, Cuttler and Welsh revisited the Lewis paper and added the dashed curve in Fig. 18 below. This was informed by their examination of the complete 65-year LSS data set, not just the first ten years that were available to Lewis. And heads up—these are logarithmic graphs, squeezed down from the top and in from the right. It's a statistician's trick that shrinks a large data range down to a single page, while the critical low-dose range is still easy to see.

Examining the complete data set, with the advantage of nearly six additional decades of science on biology, radiation, and

FIGURE 18: Cuttler and Welsh Data Set (2015)

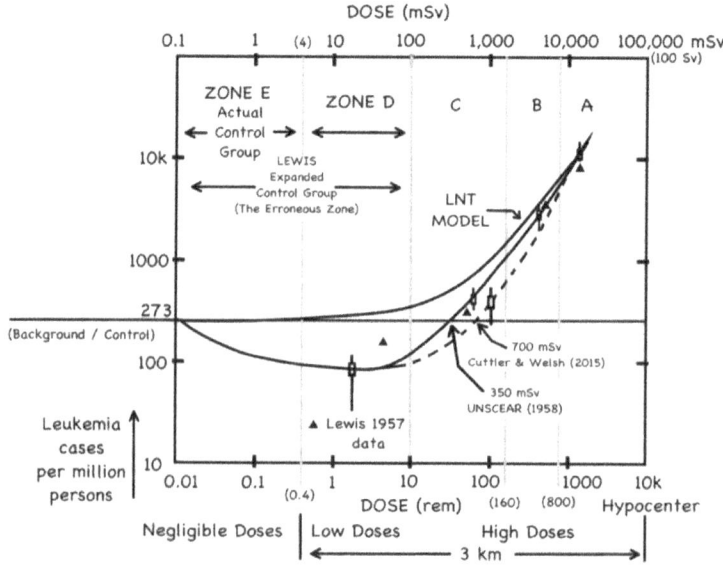

Source: https://www.researchgate.net/publication/288992343_Leukemia_and_Ionizing_
Radiation_Revisited (with apologies to Cuttler and Welch)

health effects, Cuttler and Welch adjusted the 350 mSv threshold seen in the 1958 data to 700 mSv, or 0.7 Sv, the point where their dashed curve exceeds the 273 cases-per-million leukemia rate in Zone E, representing the wider postwar population. (Note: MilliSieverts are scaled across the top of the graph, and rems are scaled across the bottom.)

Now let's go back to 1957, and see how Lewis viewed the UNSCEAR data:

FIGURE 19: Lewis Views the Data in 1957

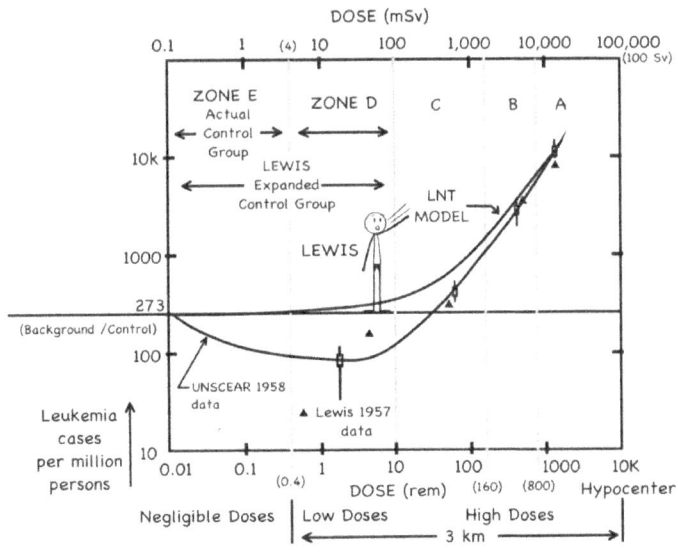

Source:
*https://www.researchgate.net/publication/288992343_Leukemia_and_Ionizing_Radiation_*
Revisited

Notice that he isn't looking at one of his own data points— the triangle below his right foot. This point shows a "reduced effect" in Zone D, with a significantly lower rate of leukemia than the background/control line he's standing on. Above the line, excess rates of leukemia can be attributed to excess radiation from the blasts; no argument there. But below the control line, a small dose of radiation from the blasts is associated with a lower rate of leukemia. This of course is contrary to LNT.

Standing at the leading edge of his Erroneous Zone (the Expanded Control Group in Fig. 19), the data looming before Lewis looks alarming, indeed. His three triangles in the high-dose range (Zones C, B, and A) conform quite well to Muller's Proportionality Rule. Seeing this, Lewis concluded that the linear response predicted by LNT for inherited mutations also applies to leukemia. If this were true (it's not), then LNT would probably apply to other cancers as well (it doesn't).

When Lewis was writing his paper in 1957, a proper reading of the Life-Span Study data would have revealed two remarkable things:

- A hormetic (adaptive) benefit may lie somewhere between 350 mSv and background dose.
- Anything below 350 mSv is apparently safe for leukemia, and probably for other cancers as well.

Only the first decade of LSS data was available to UNSCEAR, and thus to Ed Lewis and his contemporaries, while Cuttler and Welsh had fifty years of additional data on the same people and their offspring. Even so, the UNSCEAR curve and the Cuttler-Welch curve are in good agreement. Both are well away from the LNT line, and both show a substantially lower rate of leukemia in the Zone D survivors compared to the non-irradiated survivors from Zone E.

Finally, let's take a look at Figure 20 below, where everything comes together in one handy graph. When we (the "Not Lewis" guy) assume a vantage point to take in all the data, five things become abundantly clear:

- A high-dose proportional response at high doses
- A possible threshold at 700 mSv
- An established threshold at 350 mSv

- A hormetic response below 350 mSv

- LNT is BS.

FIGURE 20: Don't Be Lewis!

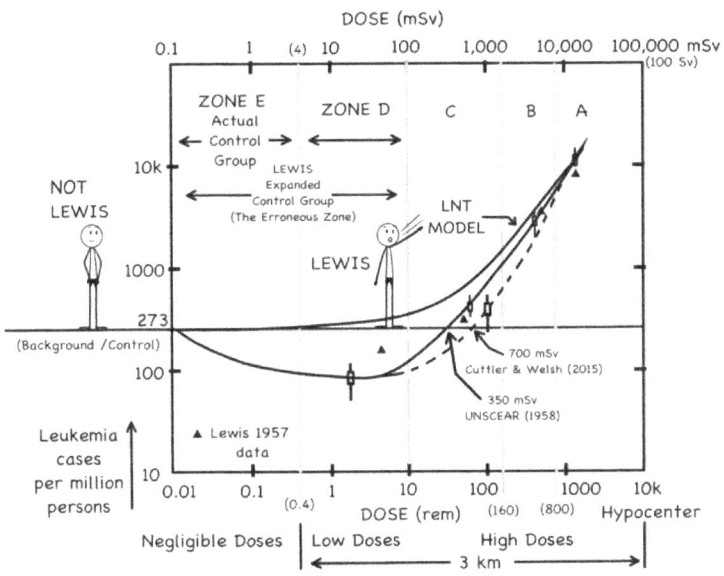

Source: https://www.researchgate.net/publication/288992343_Leukemia_and_Ionizing_Radiation_Revisited

With Zone E restored as the proper control group, Zone D can be seen for what it is—moderately-irradiated people who were closer to the blasts than the general population, but not too close, and who subsequently had lower rates of leukemia than those who were either closer to the blasts or farther away.

12.4 Discussion

As we've seen, the threshold for ionizing radiation is much higher than our regulatory agencies acknowledge. Be that as it may, getting them to acknowledge any threshold at all, even a modest 100 mSv per year, is going to be a formidable task, in spite of the fact that after nearly a century of study, no deleterious low-dose effects have been proven.

Nuclear fear being what it is, adjusting the safety standard to the 1958 threshold of 350 mSv would be a minor miracle, even though a threshold of 700 mSv is supported by the evidence. But it is reassuring to know that a regulatory adjustment to even 350 mSv would still ensure a substantial margin of safety.

To be clear, no one is suggesting we let our guard down and expose ourselves to danger. Rather, we're suggesting that all of us quit being so overly damn scrupulous about every last smidgen of ionizing radiation, to the point where we shy away from a proven clean-energy technology that can power the nation, and the planet, all by itself if need be, for literally centuries to come. Even a modest regulatory adjustment, from the current no-safe-dose assumption to just 100 mSv per year, would ensure a substantial margin of safety while significantly lowering the cost, and raising the public acceptance, of nuclear power.

To cite a few examples:

Since 2011, the Japanese government has been removing Fukushima topsoil to restore downwind croplands to one milliSievert above the background dose, while the displaced farmers fritter away their lives in "temporary" housing. The cost of this Herculean effort? About $25 billion.[10]

To put Fukushima's scary soil in perspective—the yearly background dose in Japan is 2 mSv per year, compared to 7 mSv per year in Denver. And Denver has normal rates of cancer. (See *Earth Is a Nuclear Planet*, Chapter 14.)

On another front in their ongoing battle against low-dose radiation, the Japanese government is using seawater to dilute 1.3 million tonnes of the reactor flush water they've stored since the accident. This fresh water has been filtered of all radioactive contaminants except for tritium, a slightly radioactive form of hydrogen, which can't be filtered out because hydrogen is an integral part of the water molecule. Japan's scrupulous safety standards require them to get the tritium levels down by diluting

Fukushima, Japan – Tritiated Water Storage
Credit: IAEA Imagebank (CC-BY-SA-2.0)

it with seawater, to a concentration that is fifty-three times below the WHO's recommended tritium levels for drinking water, before releasing the treated water into the ocean, far from shore and a bit at a time.

As we show in *Earth Is a Nuclear Planet*, this "Fuku water" is so benign that you could drink a half-shot every day of your life, straight from the storage tanks and prior to seawater dilution, and all it would do is match the internal radiation you're already getting from your dietary potassium. Making this water even more harmless than it already is will cost them about $3 billion.[11] And people wonder why nuclear power is expensive.

Another example: Here in the US, the land that once hosted a nuclear power plant must be restored to no more than 0.25 mSv above the background dose. While this may sound prudent to a society raised on nuclear fear, most people don't know that the background dose in the US ranges from about 1.3 mSv per year in Florida to over 9.6 mSv in South Dakota.[10] This means that a site restored to 9.85 mSv (9.6 + 0.25) in South Dakota would be 8.3 mSv above the federal standard for a decommissioned nuclear plant in Florida. This of course is absurd, prejudicial, and costs a needless fortune, while reinforcing nuclear fear and doing nothing to improve health and safety.

Unless things change, the clean energy we need to power the nation, and the planet, will continue to be passed over in favor of less feasible technologies. With Muller's no-safe-dose approach to radiation safety, we're going to have one hell of a time meeting our clean energy goals by mid-century, or by 2100 for that matter.

CHAPTER THIRTEEN
Calabrese, Cuttler, and McNutt (2015–present)

IN 2015, when Ed Calabrese first came across the Cuttler and Welsh paper, he thought the authors were "seeing something that wasn't there." [1] But the study had been peer-reviewed so he took a closer look; if it held water, it would be a major game-changer. As a seasoned toxicologist and noted hormesis researcher, Calabrese was intrigued.

He revisited several postwar studies that showed a tell-tale J-curve pattern in the leukemia data of the Life-Span Study. For example, consider Wald's 1958 paper in the journal *Science*, the third study listed in Calabrese's meta-graph (Fig. 12). It describes a classic J-curve dip: 273 cases per million down to 92 cases, and back up through 273 to 398—the same numbers cited by UNSCEAR.

Aside from the nine J-curves he plotted on his graph, Calabrese discovered something else—a telling detail in an unpublished draft of the Lewis paper, the same draft reviewed by the BEAR-I genetics panel. The draft paper cites a 1954 study by Furth and Upton[2] that found a clear safety threshold in the mice and pigs irradiated in US atomic bomb tests in Nevada. This confounding data was omitted from Lewis's published version.

Over the last several years, Ed Calabrese has assembled the long, strange tale of LNT, which includes a trove of personal correspondence between Muller and his lifelong friend and pen pal Edgar Altenburg, along with the correspondence of Curt Stern, Ernst Caspari, and others.[3] Understandably, one of the few documents Calabrese could not obtain from the Muller estate was the suicide note that Muller wrote to Altenburg, but aside from that the data trail is remarkably complete.

Reviewing these letters alongside the contemporaneous work of their respective writers and recipients, and the studies they were discussing (and at times misinterpreting) in their private correspondence, Calabrese has documented a clear and convincing case of scientific misconduct.

McNutt rejected Cuttler's request, and Cuttler forwarded their exchange to Calabrese. In August 2015, Calabrese emailed McNutt to respond to the several points she made in her reply to Cuttler, and implored her to reconsider. They had a series of exchanges that went nowhere, her last response ending with: *"Please respect that the matter is closed."*

Nevertheless, Calabrese persisted, continuing his research and eventually publishing his October 30th 2017 paper "Societal Threats from Ideologically Driven Science," which includes his exchange with Dr. McNutt.[4]

13.1 The Controversy Continues

The cell-staining technique that Barbara McClintock developed enabled her and Lewis Stadler to disprove Muller's work with visual evidence as early as the Cornell Conference in December 1931. This is the same work for which Muller received the Nobel in 1946. Since then, his LNT model of radiation risk assessment has been:

- Disproven by Snell in 1932
- Affirmed by Uphoff and Stern in 1949
- Disproven by Neel and Schull in 1956
- Adopted by BEAR-I in 1956
- Contradicted by the BMRC in 1956

- Metastasized by Lewis in 1957

- Validated by the NCRP in 1958

- Re-affirmed by BEIR-I in 1972

- Adopted by the EPA in 1975

- Disproven by Paul Selby in 1996

- Re-affirmed again by BEIR-VII in 2006

- Disproven again by Cuttler and Welsh in 2015

- Disproven yet again through extensive historical documentation by Calabrese, from 2015 to present.

A long, strange tale, indeed. At the conclusion of his eleven-hour interview with the Health Physics Society, Dr. Calabrese offered these remarks (lightly edited for clarity):

> The impacts of LNT on society are pretty pervasive, and they extend from government, to the research community, to the education community, to political frameworks, to how the media frames questions, and so forth.
>
> *It's very pervasive and it works on fear.* And that's a real issue. It's very easy to play on people's fears and to be afraid of serious things, such as birth defects or cancer. Nobody wants anything like that, so it's an easy thing to get worked up about and to blame others about.
>
> The biggest problem is that regulatory agencies have somehow disconnected themselves as scientists from understanding who we are in an evolutionary sense. *They tend to view humans and all life as victims, and as victims unable to repair and to withstand the challenges we have.*
>
> Within our biology, we have many different types of adaptive responses, and these adaptive responses are activated by low levels of stress, and many times they are up-regulated to an extent that they not only just fight off a challenge, but they can give you a benefit beyond the challenge.

The body can respond to any sort of threat, and it has to if it's going to survive, by defending itself. *Not every dose causes a toxicity, and if it causes a toxicity, the toxicity doesn't remain—the body responds and repairs itself.*

The EPA regulations, are built upon a framework of the cell and the individual as being victims, especially when *you can see in their regulations that the goal is zero exposure. But we live in a world that is anything but zero exposure.*

What's a healthy vegetable? It may be a vegetable that induces oxidative stress within your biology so that you can up-regulate the anti-inflammatory frameworks [in your body] to protect yourself. The human dynamic needs to be brought into the equation here; biology should govern the regulation here, and it's not doing so. And that to me has been a problem.

If you look at the EPA's acceptable level of exposure [for drinking water] and if you count the number of molecules [of a regulated carcinogen] in a liter of [that] water, chemical carcinogens are present in drinking water at about one trillion molecules per liter, up to perhaps a quadrillion molecules per liter. *You could [therefore] be exposed to trillions of [toxic] molecules every single day, and it [would be] below the [EPA-estimated] risk of one in a million.*

You don't see any action [i.e., biological effects] in animal models until you get up into the many trillions of molecules [of a particular toxin per liter] per day. So something is going on biologically within us that *it takes a certain level of exposure before anything happens.*

I just think the EPA has forgotten their biology, and forgot that we have a tremendous capacity to repair. There is a wealth of scientific data in many different biological models, from microorganisms to humans, to show that J-shaped dose responses occur.

We have inherited a distasteful history by leaders in our field that has brainwashed us into believing their flawed and ideological science. It's corrupted our regulatory practice.[5] [*emphasis added*]

In 2016, Marcia McNutt, the editor of *Science*, was elected president of NAS, now called NASEM (National Academies of Sciences, Engineering and Medicine). In June 2022, NASEM issued a 300-page proposal calling for a multi-decade/multi-disciplinary inquiry, to be conducted by several government agencies at an estimated cost of about $100 million a year, to explore the potentially harmful effects of low-dose ionizing radiation.[6] Curiously, the proposal makes no mention of hormesis, and cites just one paper skeptical of LNT. And no one on the panel that issued the paper is an LNT critic.

While the NASEM proposal does acknowledge a strong and growing pushback against LNT, it also points to increasing evidence of possible harm from non-cancerous effects of low-dose radiation, such as cardiovascular and brain function issues. While nothing definitive has been found, there are hints that warrant further study. Fair enough, and further research is entirely appropriate. But assuming that harm exists until proven otherwise violates the scientific principle of letting evidence lead the way. Research must always proceed from a "null hypothesis"—an initial assumption that no effects exist (null) until and unless it can be demonstrated otherwise.

As they say in the world of construction, "ASSUME makes an A-S-S out of U and ME." And after sixty-plus years of nuclear power generating twenty-seven trillion kilowatt-hours for the US grid,[7] applying the Precautionary Principle to commercial nuclear power is assumptuous in the extreme. It's an outdated, fear-based approach to public health policy on ionizing radiation that has nothing to do with science, or the scientific method.

> I would rather have questions that can't be answered, than answers that can't be questioned.
>
> —RICHARD FEYNMAN, Physicist

The Precautionary Principle can be an entirely valid approach when it comes to new and untested technology with uncertain consequences. But after all this time and after twenty-seven tril-lion carbon-free kilowatts, an obsessive focus on absolute safety is sophistry, not principle, because it slyly assumes the ability to prove a negative at the cellular level, in isolation from any other stressor.

And once you buy into that squirrel cage, nuclear science ends and nuclear fear begins. It's like the ancient mapmakers warning of *"Here Be Dragons!"* in the oceanic fringes of the New World. They of course had no idea, but when medieval imaginations fill a vacuum it is seldom done with dolphins and rainbows.

Even today's cutting-edge science cannot discern every spe-cific cause-and-effect relationship in the low-dose range, partly due to the fact that there is just so much activity on the cellular level—the animation in endnote 10 of Chapter 7 is a tiny frac-tion of what goes on inside a cell. Living stuff is far too busy and ever-changing to permit such fine discrimination; separating the effects of low-dose radiation from the effects of other stressors often cannot be done.

For example, the biological effects of oxygen-induced free rad-icals; of micro-doses of toxins; and low doses of ionizing radiation are all essentially the same, one to the other. They all break im-portant biochemical bonds which in some cases may not be ade-quately repaired or compensated. Determining which agent caused which effect is often impossible, even with today's instruments.

Even so, we do know that nearly all damage from ionizing radi-ation is routinely self-repaired by a healthy cell. And how do we know this? Because once again, here we are, having successfully evolved over billions of years from rugged colonies of single-cell organisms—pond scum, if you will—on a far more radioactive planet than we now enjoy. The empirical evidence is hardly in doubt.

Despite the assurances of neutrality in the preface of the NASEM proposal, the paper's overall approach is biased. There is a brief mention that while low-dose exposures "may yield an individual or societal benefit, they may also adversely affect human health." [8]

Well, okay . . . and you may also get hit by a meteor at any moment. Don't laugh; it could happen—dismiss the possibility at your peril. So, just to be safe, shouldn't you install an inch of steel on your roof? How about two inches to be extra super-duper safe, because you can never be too sure?

Actually, you can, and the over-regulation that results does nothing to enhance survival. And as the years roll on, the consequences of not using clean power on a massive scale at every opportunity will far outweigh the minimal risks of commercial nuclear power. Most people would be astonished to learn just how safe the technology actually is; we've laid out the plain-language facts in *Earth Is a Nuclear Planet*.

The authors of the NASEM proposal do go out of their way to assure the reader that those who dispute LNT have been heard loud and clear. But aside from these hopeful opening remarks, the paper's overall position is hard to miss. To paraphrase and summarize:

> A thorough study of low-dose radiation will unfortunately require decades of extensive and expensive lab work, conducted by multiple federal agencies. Until such work is complete, please be advised:
>
> - The LNT model shall remain in force.
> - All radiation is assumed to be harmful—Here Be Dragons!
> - After seventy years, the Precautionary Principle is still appropriate when it comes to something as scary as low-dose radiation.
> - This being the case, ALARA (As Low As Reasonably Achievable) is prudent and appropriate policy.
> - The burden of proof is incumbent upon those who disagree.

The fact that the original work on LNT was funded by a fossil-fuel dynasty that had, and still has, trillions of very good reasons to restrict nuclear power just makes it all that much worse. To give you an idea of the stakes involved: The estimated value of the world's recoverable petroleum reserves—what the industry knows is there, and what they know they can extract, is about 100 trillion dollars.[8] And that was before the discovery off the Guyana coast.

As Calabrese has documented, Marcia McNutt, the current head of NASEM, has no further interest in discussing the matter. Instead, NASEM is kicking the LNT can down the road by floating a multi-decade, multi-billion-dollar proposal. And if it ever gets funded, and completed, circle back to them in a few decades. Until then, kindly review the above talking points . . .

Sorry, but we don't have a few decades.

Hermann Muller developed his model of radiation risk assessment nearly a century ago, before low levels of radiation could even be accurately detected, measured, and delivered, and his conclusions have long since been proven wrong. And yet here we are, in a warming, carbon-choked world,[9] with far too many people still freaked out about nuclear power—even though you get more radiation from eating a banana than you'd get from living near a nuclear power plant for an entire year.[10]

Behold the power of nuclear fear.

The bottleneck is, and has always been, Muller's Linear No-Threshold model of radiation risk assessment, a century-old hypothesis with no solid science to back it up.

Dr. McNutt, NASEM, EPA, NCRP, DoE, NRC, *et alpha-beta*, please respect that the matter is far from closed.

CHAPTER FOURTEEN
A Way Forward: The Optimization Model of Radiation Risk Assessment

As Ed Calabrese points out, the subject of radiation risk can arouse such strong passions that it can be difficult for people to agree on much of anything. As he readily admits, "*I recognize that I'm not going to convert the world to believe the way I believe, and someone's probably not going to convert me, either.*"[1]

To break the deadlock, he offers a common-sense compromise to which even those in entrenched positions can agree, without feeling as if they've lost the good fight for public safety and abandoned their principles, precautionary or otherwise.

The basic idea is to combine the three radiation-response models depicted in Figure 11 (linear, threshold, and hormetic) into a single blended model that should satisfy all entrenched positions in the radiation safety debate. In his Optimization Model, an agreed-upon rate of risk could be arbitrarily set at, for example, one excess cancer in 10,000 persons over the course of a lifetime.

This is by no means a reckless proposal—cancer already occurs in approximately one out of every three people in modern society.[2] This means that an additional risk of one more case of cancer in 10,000 people (100 cases per 1,000,000) is statistically equivalent to assuming that there is no radiation safety threshold, and that a linear dose-response exists all the way down to zero dose / zero effect.

In other words, allowing for one additional case of cancer in 10,000 is tantamount to accepting LNT, without having to get all hyper-scrupulous by striving to Go As Low as Ridiculously Achievable (ALARA).

Why? Because as Calabrese explains in his Health Physics Society interview, such a minuscule degree of additional risk is literally "beyond the detection capacity of studies . . . A risk of one in 10,000, or one in a million, is simply an artificial construct—it's not provable." [3] From the interview:

> If you take an animal study and you assume that the [toxic] agent acts by a threshold and not by a linear dose-response, and you do the methodology the way the EPA would do it [i.e., by assuming there is no safety threshold], the number I come up with [in my model] is nearly identical to a [lifetime] risk of 1 in 10,000.
>
> In the hormetic or J-shaped dose response model I work with, *just about every chemical toxin shows a benefit at low doses.* The nadir of that beneficial curve [i.e., the lowest, or maximum beneficial, point] matches the other two [models.] [*emphasis added*]

Here is Figure 11 again, showing the three models:

Threshold Dose-Response Models

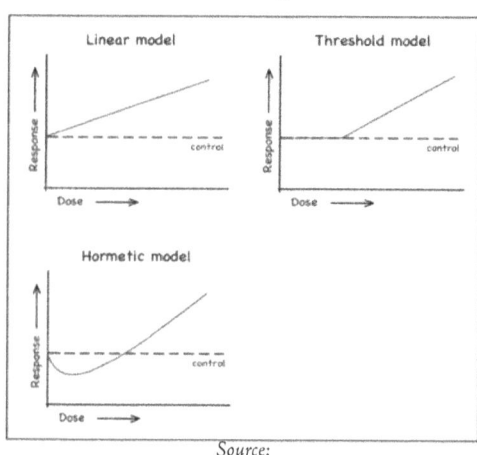

Source:
http://wwwtoxicdogy.org/guys/RSESWebinar10415.pdf

This is what they look like blended together:

FIGURE 21: Blending the Models

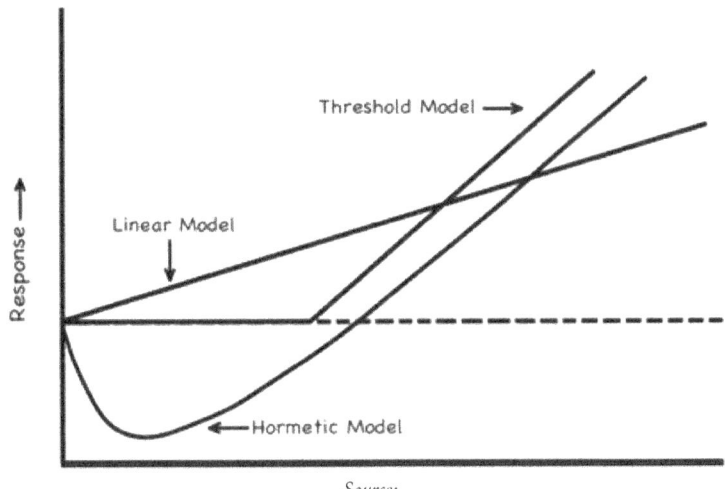

Source:
https://www.toxicology.org/groups/ss/RSES/doc/RASS-RSESS_Webinar_101415.pdf

And this is Dr. Calabrese's compromise:

FIGURE 22: The Risk Optimization Model (Calabrese 2021)

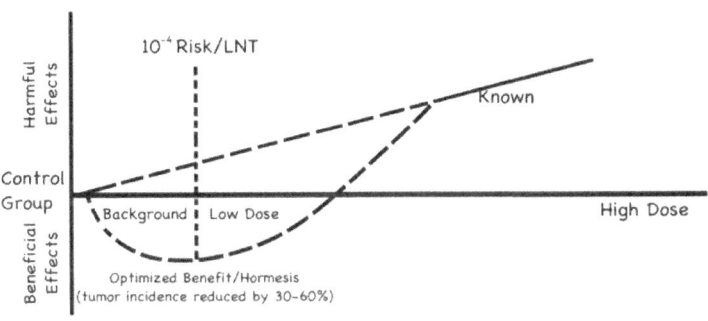

Source:
http://hps.org/hpspublications/historylnt/episodeguide.html
(Note: 10-4 means one case in ten thousand, or 100 cases per million)

Calabrese goes on to explain:

> If I'm totally wrong, and LNT is correct, then the amount of adverse effects on the population would obviously increase, but it would still be relatively modest—one in 10,000 compared to a normal expectation [for cancer] of one in three.
>
> If I'm right [about hormesis], it would be a tremendous benefit to society, far greater than any risk. You would be reducing the occurrence of certain types of cancers by about 30 to 40 percent [by allowing for low-dose exposures that prompt an adaptive response].
>
> Over time, studies will say that results are favoring [either] this model or the other—that's what the science will tell you in the future. What the [Optimization Model] would do [for us right now] is it would break [people] away from the extreme positions we see in society [which result in] extreme costs for clean-ups, which are things that discourage new technologies from being developed.
>
> [The Optimization Model] would result in a more workable, common-sensical, and at the same time a more intellectually honest [approach]. I'm coming to the table saying, 'I think I'm right, but I could be wrong.' And you could come to the table saying, 'Well, I think *I'm* right, but I could be wrong.' But we could now go forward and try to see which model actually gives us the best data. This would allow us to get away from the crippling extreme that LNT has in the population, where people are afraid of ephemeral risk.
>
> And even if somebody were to claim there's a risk of one in 10,000, there would be no way to prove it, because *it's beyond the detection capacity of studies to do that.* A risk of one in 10,000, or one hundred in a million, is simply an artificial construct—*it's not provable.*
>
> I'm asking people to accept a higher risk, [but a risk] that could never be proven epidemiologically [one way or the other]. That's still a very conservative position to be in.[4] [*emphasis added*]

For the last seven decades, Hermann Muller's no-safe-dose model of radiation risk assessment has demonized nuclear power, the world's safest and best clean-energy solution. We hope this little book of ours helps Dr. Calabrese get his ideas out to a wider audience. His work is a stake through the heart of LNT, and we're only too happy to help hammer it home.

SUPPLEMENT

Sex and the Single Gamete: A Crash Course in Reproduction, Sexual and Otherwise

What was known in the 1950s about cellular life was considerable, though most of it was limited to direct observation. As far back as the 1870s, mitosis, or asexual cell division, had been observed in both reproductive (germ line) cells and somatic (body) cells.[1]

It was also known that, unlike reproductive cells, somatic cells further their genetic line by dividing into an identical pair of new cells, without hooking up with another member of the same species to produce offspring—mitosis is how your skin makes new skin.

It was also known that reproductive cells go through a two-step process called meiosis. But the details of DNA replication, central to both mitosis and meiosis, weren't described until the 1960s, and the molecular mechanisms of either process weren't detailed until the 1990s.[2]

When Muller and Lewis were exploring radiation risk, and when the world's geneticists accepted their assumptions, and when those assumptions shaped our perceptions of safety and risk and became enshrined in our federal regulations, none of them knew how any of this actually worked.

If biologists in Muller's time had understood the intricacies of gamete fusion, DNA replication, cell division, and especially cellular self-repair, they would have known right away that Muller was wrong, and he would have known as well. But back then, exactly how cells replicate was as magical and mysterious as the sorcerer's broom in *Fantasia*.[3]

Mitosis

In mitosis,[4] a cell does not "give birth" to another cell. It first makes a copy of its own DNA, then divides into an identical pair of brand-new cells, each with a full set of DNA "blueprints." These new cells repeat the process, copying their DNA and then dividing into more cells, much like the sorcerer's broom.

The four steps of mitosis (asexual cell division) are Prophase, Metaphase, Anaphase, and Telophase, or PMAT. Meiosis (sexual cell division) goes through these steps twice (see below).

The plump X and Y chromosomes we're all familiar with are actually long "chromatin" DNA molecules that have coiled themselves into compact, organized bundles called "chromatids." Individual strands of chromatin, the chromatid bundles they curl into, and the X and Y structures they form when they link at their centromeres (see below), all fall under the term "chromosome."

Before mitosis or meiosis begins, the strands of chromatin in a cell's nucleus look like a tangle of spaghetti (not shown in Figure S-1; we've all seen spaghetti). In a set-up step called Interphase, the chromatin strands are replicated and proofread for errors.[5] With that done, a cell duplicates its DNA by 'unzipping' its chromatin and adding new bases to each half. The result is a matched set of chromatid strands.

In the middle of each strand is a "primary constriction site" called a centromere.[6] After replication and proofreading, the strands (both the originals and the copies) shorten, thicken, and coil into plump, individual chromatid bundles (Figure S-2, top left). Then the matched pairs (an original and its copy) join at their centromeres like the nut and bolt that hold a pair of pliers together (Figure S-2, lower left).

FIGURE S-1: DNA Replication

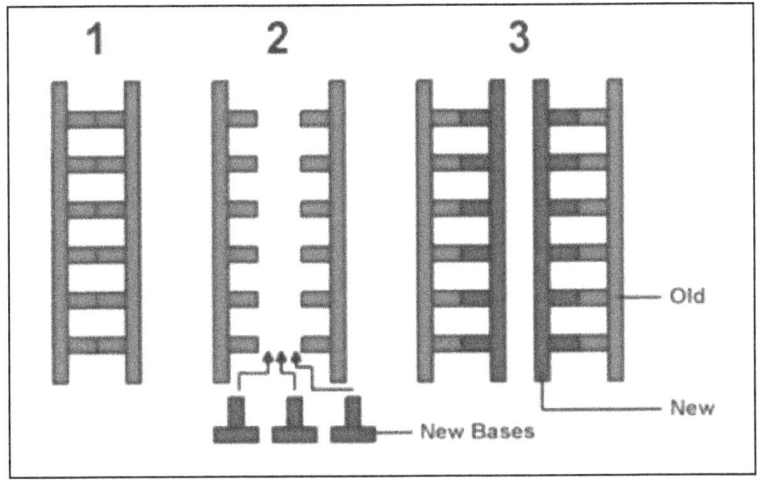

FIGURE S-2

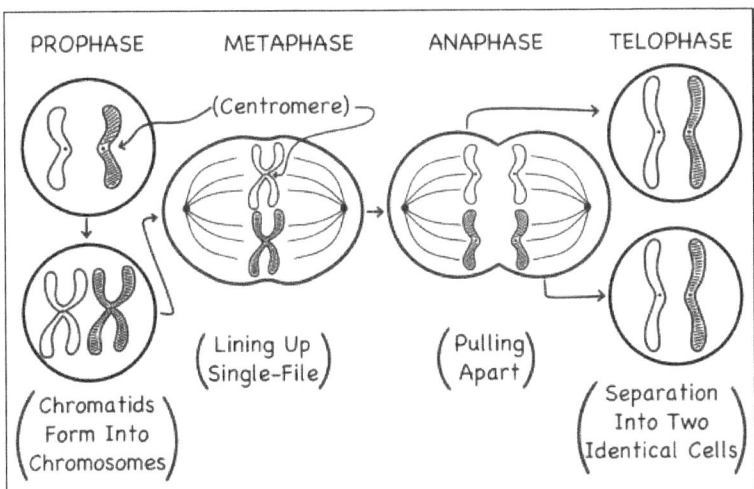

Credit: By the author

In Metaphase, these "sister chromatids" line up in the middle of the cell. In Anaphase, the pairs are pulled apart as the cell begins to divide. In Telophase, cell division is complete, resulting in two new cells, each with a full set of DNA.

After the excitement is over, the plump chromatids in these new cells uncoil into long, individual strands of chromatin, looking once more like a tangle of spaghetti.

Since every somatic cell has a complete set of DNA, there is no need to fuse with a cell from another organism of the opposite sex to produce new somatic cells. They just make copies of themselves, all by themselves, without a mate—no muss, no fuss.

Asexual cell division is an ongoing, lifelong process, from the simplest pond scum to the most complex primate. As we saw in Chapter 7, mitosis in humans occurs about one trillion times per day. And the self-repair within these cells is a whole other blizzard of activity: With 10,000 repairs per cell per day, and with 32 trillion cells in the average adult human, our DNA is being repaired about 3.7 trillion times per second.

With this stupendous amount of non-stop activity, the mechanisms have to be fundamental, resilient, and virtually foolproof. If cells were anywhere near as delicate as Muller and Lewis thought, we never would have gotten this far.

Reproductive Cells Do It Differently

More properly called gametes, reproductive cells are the specialized somatic cells in advanced (sexually reproducing) organisms that have the ability to undergo meiosis, a two-step strategy for extending the organism's genetic line.[7]

Like any other somatic cell, a gamete starts out with a full set of DNA. This means that immature gametes can, and often do, multiply through plain old asexual mitosis, just like any other somatic cell. This is commonly known as "growth."

And just like regular somatic cells, immature reproductive cells start proliferating long before birth. In fact, at twelve weeks, a female embryo starts producing all the eggs she will need for her entire life, finishing the process months before she is even born.

Years later, at the onset of puberty, and for the next forty or so years after that,[8] a small portion of the female's immature gametes (eggs) will periodically undergo a two-step meiosis process, something that regular somatic cells cannot do. Meiosis is what turns an immature gamete, with a full set of DNA, into four mature gametes with a half-set of DNA each (see below).

Similarly, a male embryo will produce immature sperm cells (spermatagonia) long before birth, and will continue producing immature sperm after birth, through childhood, into adulthood, and hopefully into old age. But at the onset of puberty, and unlike the periodic gamete maturation in the female, about 30% of a male's spermatagonia at any one time are maturing into spermatozoa (see Chapter 8). And just like in the female, the two-step meiosis process turns a maturing spermatazoa cell into four mature gametes, each with a half-set of DNA. And then, they grow a tail.

That's the unique thing about reproductive cells (aside from males growing tails)—when gametes mature, they only have *one-half* of their original set of DNA, and this is one reason why mating is always a bit of a crap shoot. Since any gamete needs a complete set of DNA to develop into a new, individual organism, they must fuse (mate) with a gamete of the opposite sex so the two of them can zip together a full set of blueprints. This is commonly known as "conception."

FIGURE S-3: DNA "Zipping"

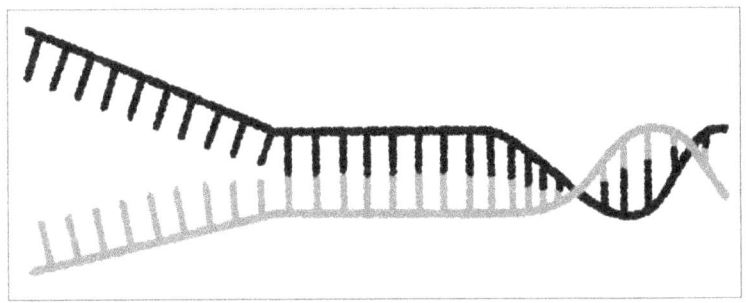

The fertilized egg that results is a somatic cell with a complete set of DNA—that's what makes it a somatic cell. It's the start of a new and line of somatic cells that can develop into a unique,

FIGURE S-4: Meiosis—Prophase I

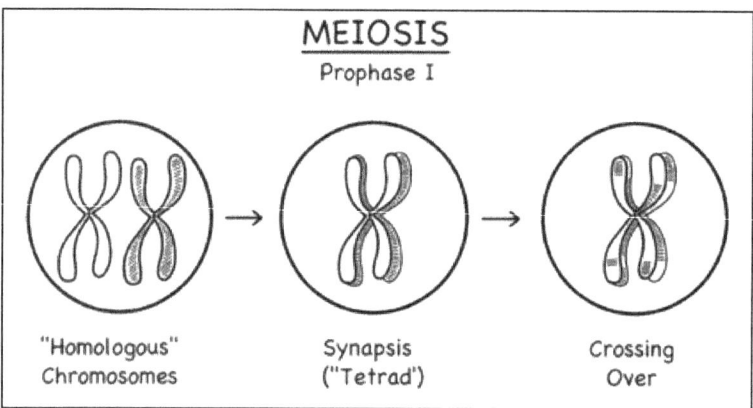

Credit: By the author.

individual organism, with a full set of blended DNA assembled from the two half-codes of the male and female gametes that fused during conception. This is how complex life forms extend their genetic lines (hopefully, just the good parts) into the next generation. And that's the rest of the story of the birds and the bees.

Meiosis

Gamete formation takes place through meiosis, which has two PMAT sequences.[9] Let's look at these in sequence.

In Prophase I, the first step of the first meiosis sequence (cleverly called Meiosis I), the chromatid sister pairs in a reproductive cell will link together at their centromeres (Figure S-4, left side), just like any other somatic cell does in the first phase of mitosis.

But then, something special called "synapsis" takes place (Figure S-4, center). This is the first evolutionary step beyond mitosis, in which the gamete's similar, or "homologous," chromosomes—one inherited from the mom and one inherited from the dad—partner up to form "tetrads."

This brings together the same kind of genes on each chromosome. For example, the gene for hair color on the mom chromosome is snuggled up beside the gene for hair color on the dad chromosome, along with dozens of other unique features from each parent, everything from bad teeth to a predisposition for diabetes, and all these various traits are dancing cheek-to-cheek.

Then another special something called "crossing over" takes place (Figure S-4, right side). This is where the two chromosomes in a tetrad exchange sections of DNA with each other—not by copying each other's DNA, but by actually trading physical sections of DNA with each other. So now the dad chromosome, with his bad teeth and predisposition for diabetes, has mom's hair color, and the mom chromosome, with her dimples and big feet, has dad's hair color. The variability and occasional surprises that result from this genetic mash-up provide a greater range of traits that can help the species adapt to a changing environment.

Figure S-5 shows the next three steps, the M-A-T of PMAT-I. (To keep things simple, only one tetrad is shown.) In Metaphase I, the tetrads line up in the middle of the cell, just like chromosomes do in mitosis.

FIGURE S-5: Meiosis—the M-A-T of PMAT I
Metaphase I, Anaphase I, and Telophase

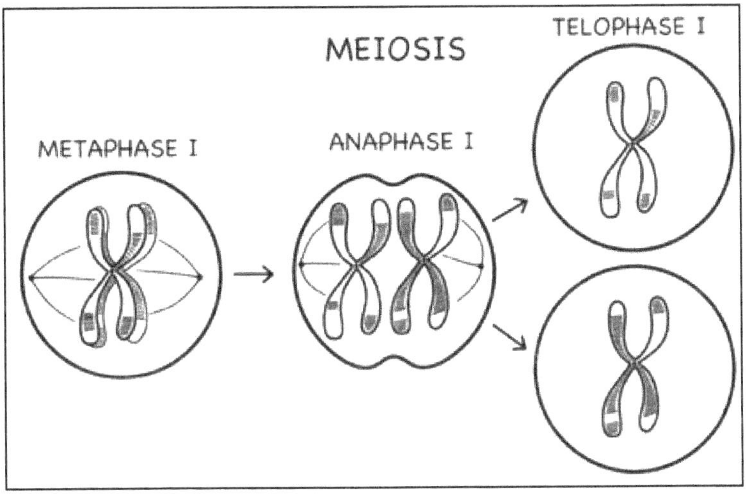

Credit: By the author

In Anaphase I, as cell division starts, the tetrads are pulled apart, their crossed-over chromosomes going to the opposite sides of the cell. In Telophase I, cell division is complete and two new cells are formed, each one sporting their own unique blend of inherited mom and dad chromosomes, from which four mature gametes will be made in the next meiosis sequence.

In Meiosis II, our two new cells will go through the PMAT sequence again, starting this time with the second half of Prophase (Figure S-2, top left). That's because their chromatin strands have already bundled themselves into plump chromatids, courtesy of Meiosis I.

The other steps in PMAT-II (Fig. S-6) will turn our two new cells (immature gametes with a full set of blended DNA) into four mature gametes, each with a half-set of blended DNA.

FIGURE S-6: Meiosis PMAT II

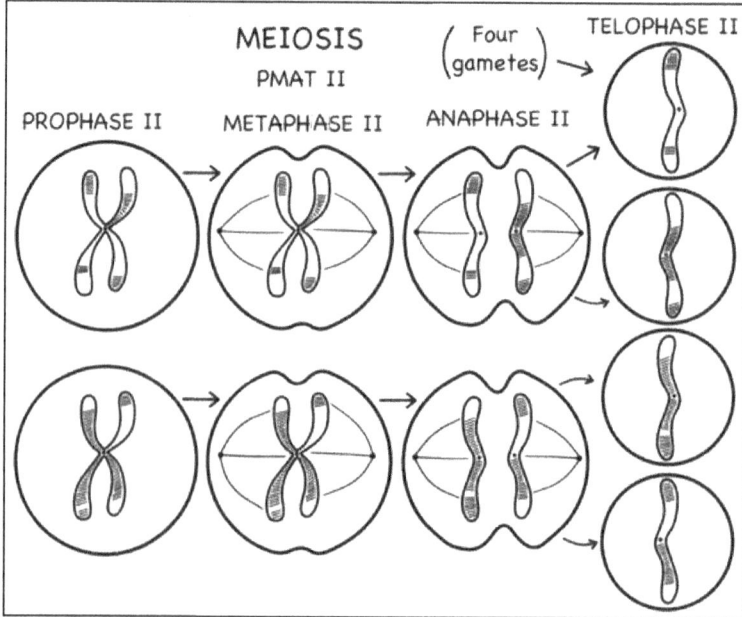

Credit: By the author

In Metaphase II, the blended chromosomes line up in the middle of the two cells, which have begun to divide. In Anaphase II, the chromosomes are pulled apart as the cells continue to divide. In Telophase II cell division is complete, resulting in four mature gametes, each with a half-set of DNA and ready to fuse with a gamete of the opposite sex. During the act of fusion (see Chapter 8 for all the juicy details), the two gametes melt into one as they zip together a full set of DNA. (Actually, it's the male gamete cell body that melts away after depositing its DNA into the female gamete.) To summarize:

- In mitosis, a DNA molecule unzips within the nucleus of a somatic cell, and the chemistry in the nucleus builds each DNA half into two separate but identical and complete DNA molecules. When the cell splits in two, each new cell has a full set of DNA. (This is how your skin makes new skin.)

- In meiosis, a mature male gamete fertilizes a mature female gamete. The female assembles a full and unique set of DNA from the two DNA halves they each brought to the party. The fertilized egg that results is a new somatic cell with a complete set of DNA, which can develop into a separate and unique organism.

And last but not least:

- Mature sperm have no repair mechanisms. All repairs are done by the female during and after conception.

Here's a handy infographic that puts it all together.

FIGURE S-7: Summary of Mitosis and Meiosis

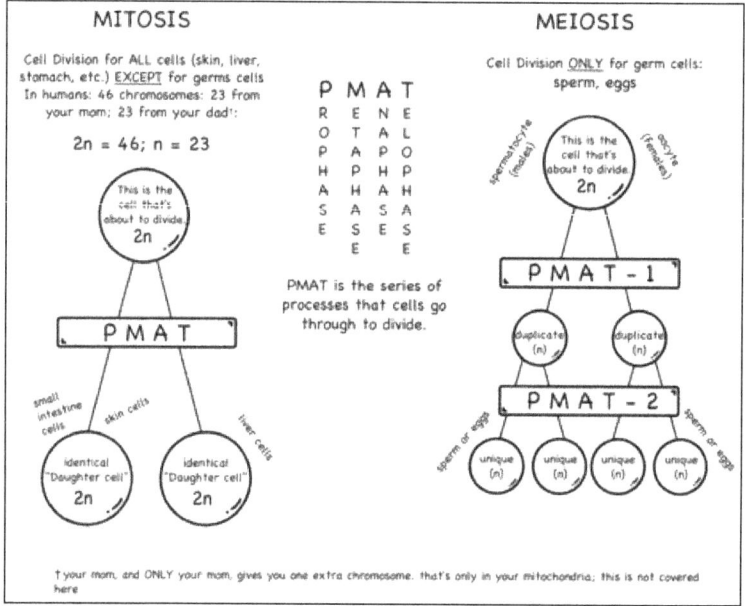

Credit: Stephen A. Boyd

Endnotes

Dedication

1. Oversight of the Environmental Protection Agency's Implementation of Sound and Transparent Science in Regulation (3 October 2018) *https://www.epw.senate.gov/public/index.cfm/2018/10/oversight-of-the-environmental-protection-agency-s-implementation-of-sound-and-transparent-science-in-regulation*

Introduction

1. The end of each episode of the Health Physic Society interview cites papers that relate to that portion of the interview, most of them written by Dr. Calabrese. The links below will open these articles, many of which have restricted access on a regular Google search. In reverse chronological order:

Recent discoveries on the historical foundations of cancer risk assessment: Shedding light on the limits of LNT (31 May 2024) *https://www.sciencedirect.com/science/article/abs/pii/S0048969724038233*

Muller and mutations: mouse study of George Snell (a postdoc of Muller) fails to confirm Muller's fruit fly findings, and Muller fails to cite Snell's findings (4 April 2024) *https://link.springer.com/article/10.1007/s00204-024-03718-1*

Cancer risk assessment, its wretched history, and what it means for public health (7 March 2024) *https://www.tandfonline.com/doi/full/10.1080/15459624.2024.2311300*

Ethical Issues in the US 1956 National Academy of Sciences BEAR-I Genetics Panel Report to the Public—(27 May 2022)
https://local.ans.org/ne/wp-content/uploads/2022/07/HP Ethical_Issues_ in_the_ US_1956_National_Academy_of.28.pdf

What would become of nuclear risk if governments changed their regulations to recognize the evidence of radiation's beneficial health effects for exposures that are below the thresholds for detrimental effects? (December 2021)
https://www.ncbi.nlm.nih.gov/pmc/articles/PMC8647278/

LNT and Cancer Risk Assessment: Its Flawed Foundations Part 1: Radiation and Leukemia: Where LNT Began (18 Mar 2021)
https://sci-hub.se/10.1016/j.envres.2021.111025

LNT and Cancer Risk Assessment: Its flawed Foundations, Part 2: How Unsound LNT Science Became Accepted (18 March 2021)
https://sci-hub.se/10.1016/j.envres.2021.111041

Ethical Failings: The Problematic History of Cancer Risk Assessment (5 December 2020)
https://sci-hub.se/10.1016/j.envres.2020.110582

The Muller-Neel Dispute and the Fate of Cancer Risk Assessment (15 July 2020)
https://sci-hub.se/10.1016/j.envres.2020.109961

Muller's Nobel Prize Data: Getting the Dose Wrong and Its Significance (8 June 2019)
https://sci-hub.se/https://pubmed.ncbi.nlm.nih.gov/31228809/

EPA Adopts LNT: New Historical Perspectives (15 May 2019)
https://scihub.se/https://www.sciencedirect.com/science/article/abs/pii/S0009279719307033

The EPA Cancer Risk Assessment Default Model Proposal: Moving Away from the LNT (July 2018)
https://sci-hub.se/10.1177/1559325818789840

The EPA Cancer Risk Assessment Default Model Proposal: Moving Away from the LNT (12 June 2018)
https://sci-hub.se/10.1177/1559325818789840

Was Muller's 1946 Nobel Prize Research for Radiation-induced Gene Mutations Peer-reviewed? (6 June 2018)
https://peh-med.biomedcentral.com/articles/10.1186/s13010-018-0060-5

From Muller to Mechanism: How LNT Became the Default Model for Cancer Risk Assessment (22 May 2018)
https://sci-hub.se/10.1016/j.envpol.2018.05.051

Societal Threats from Ideologically Driven Science (30 October 2017)
http://www.scientificintegrityinstitute.org/AQEJC121517.pdf

Flaws in the LNT Single-Hit Model for Cancer Risk: An Historical Assessment (14 July 2017)
https://radiationeffects.org/wp-content/uploads/2017/07/Calabrese-2017_Flaws-LNT-single-hit-historical-assessment.pdf

The Threshold vs LNT Showdown: Dose Rate Findings Exposed Flaws in the LNT Model Part 1: The Russell-Muller Debate (10 December 2016)
https://sci-hub.se/10.1016/j.envres.2016.12.006

The Threshold vs LNT Showdown: Dose Rate Findings Exposed Flaws in the LNT Model Part 2: How a Mistake Led BEIR I to Adopt LNT (30 November 2016)
https://sci-hub.se/10.1016/j.envres.2016.11.024

The Integration of LNT and Hormesis for Cancer Risk Assessment Optimizes Public Health Protection (17 August 2015)
https://sci-hub.se/10.1097/hp.0000000000000382

On the Origins of the Linear No-Threshold (LNT) Dogma by Means of Untruths, Artful Dodges, and Blind Faith (17 July 2015)
https://scihub.se/https://www.sciencedirect.com/science/article/abs/pii/S0013935115300311

An Abuse of Risk Assessment: How Regulatory Agencies Improperly Adopted LNT for Cancer Risk Assessment (6 January 2015)
https://sci-hub.se/10.1007/s00204-015-1454-4

Cancer Risk Assessment Foundation Unraveling: New Historical Evidence Reveals that the US National Academy of Sciences (UN NAS), Biological Effects of Atomic Radiation (BEAR) Committee Genetics Panel Falsified the Research Record to Promote Acceptance of the LNT (6 January 2015)
https://sci-hub.se/10.1007/s00204-015-1455-3

LNTgate: How Scientific Misconduct by the U.S. NAS led to Governments Adopting LNT for Cancer Risk Assessment (6 January 2015)
https://www.sciencedirect.com/science/article/abs/pii/S0013935116301219

The Genetics Panel of the NAS BEAR-I Committee (1956): Epistolary Evidence Suggests Self-interest May Have Prompted an Exaggeration of Radiation Risks that Led to the Adoption of the LNT Cancer Risk Assessment Model (4 July 2014)
https://sci-hub.se/10.1007/s00204-014-1306-7

How the US National Academy of Sciences Misled the World Community on Cancer Risk Assessment: New Findings Challenge Historical Foundations of the Linear Dose Response (4 August 2013)
https://sci-hub.se/10.1007/s00204-013-1105-6

Origin of the Linearity No Threshold (LNT) Dose-Response Concept (11 July 2013)
https://sci-hub.se/10.1007/s00204-013-1104-7

Muller's Nobel Prize Lecture: When Ideology Prevailed Over Science (13 December 2011)
https://sci-hub.se/https://pubmed.ncbi.nlm.nih.gov/22166484/

Toxicology Rewrites Its History and Rethinks Its Future: Giving Equal Focus to Both Harmful and Beneficial Effects (19 September 2011)
https://setac.onlinelibrary.wiley.com/doi/full/10.1002/etc.687

Key Studies Used to Support Cancer Risk Assessment Questioned (22 July 2011)
https://www.junkscience.com/wp-content/uploads/2011/09/calabrese-stern-paper.pdf

Muller's Nobel Lecture on Dose-Response for Ionizing Radiation: Ideology or Science? (13 April 2011)
https://sci-hub.se/10.1007/s00204-011-0728-8

The Road to Linearity: Why Linearity at Low Doses Became the Basis for Carcinogen Risk Assessment (27 February 2009)
https://sci-hub.se/https://scihub.ru/https://link.springer.com/article/10.1007/s00204-009-0412-4

Cover Up and Cancer Risk Assessment: Prominent US Scientists Suppressed Evidence to Promote Adoption of LNT
https://www.xylenepower.com/CalabreseSelby2022_Cover%20up%20and%20cancer%20risk%20assessment%20to%20promote%20adoption%20of%20LNT.pdf

2. The Health Physics Society interviews of Ed Calabrese (11 hours / 22 Episodes)
http://hps.org/hpspublications/historylnt/episodeguide.html

1. Hermann J. Muller (1927–1932)

1. Artificial Transmutation of the Gene—H.J. Muller. Science 22 July 1927
https://www.science.org/doi/10.1126/science.66.1699.84

Hermann Joseph Muller's Study of X-rays as a Mutagen, (1926–1927) by Kevin M. Gleason
https://embryo.asu.edu/pages/hermann-joseph-Mullers-study-x-rays-mutagen-1926-1927

2. *https://en.wikipedia.org/wiki/Thomas_Hunt_Morgan*

Also see para. 7:
https://www.britannica.com/biography/Thomas-Hunt-Morgan

3. See Episode 2 @ 13:30.
http://hps.org/hpspublications/historylnt/episodeguide.html

4. See Episode 4 @ 2:34.
http://hps.org/hpspublications/historylnt/episodeguide.html

Flaws in the LNT Single-hit Model for Cancer Risk: An Historical Assessment
https://radiationeffects.org/wp-content/uploads/2017/07/Calabrese-2017_Flaws-LNT-single-hit-historical-assessment.pdf
See footnotes frame 2 / page 774.

5. Fifth International Annual Conference of Genetics
https://www.nature.com/articles/124295a0

6. *https://en.wikipedia.org/wiki/Gilbert_N._Lewis*

7. See Episode 2 @ 3:30 to 6:30, and Episode 3 @ 1:00.
http://hps.org/hpspublications/historylnt/episodeguide.html
The Linear No-Threshold (LNT) Dose Response Model: A Comprehensive Assessment of Its Historical and Scientific Foundations
https://sci-hub.se/10.1016/j.cbi.2018.11.020
See Section 2: LNT and Biological Evolution.

Natural Reactivity and the Origin of the Species (Gilbert, Owen 1928)
https://www.nature.com/articles/121673a0

8. See Episode 2 @ 9:20.
http://hps.org/hpspublications/historylnt/episodeguide.html

9. See Episode 3 @ 1:34.
http://hps.org/hpspublications/historylnt/episodeguide.html

Evidence that Natural Radioactivity Is Inadequate to Explain the Frequency of "Natural" Mutations—(Hermann Muller, Mott-Smith 1930)
https://sci-hub.se/10.2307/85359

Muller's Nobel Prize Data: Getting the Dose Wrong and Its Significance (Calabrese)
https://sci-hub.se/https://pubmed.ncbi.nlm.nih.gov/31228809/

10. *https://ricehistorycorner.com/2017/12/11/edgar-altenburg-1888-1967/*

11. See Episode 4 @ 5:00.
http://hps.org/hpspublications/historylnt/episodeguide.html

12. See Episode 2 @ 12:12.
http://hps.org/hpspublications/historylnt/episodeguide.html

Muller administered 2,700 mSv in 3.5 minutes.
2.7 Sv = 2700E-3 Sv ÷ 3.5 min × 60 min /hr × 8760 hr /yr = 405E3 Sv /yr.
Average background dose in North America is about 3 mSv per year, or 3E-3 Sv /yr.
Ratio of the two doses = 405E3 Sv /yr ÷ 3E-3 Sv /yr = 135E6.
In other words, the dose rate Muller applied was 135 million times greater than the average background dose-rate found on planet Earth.

13. See Episode 3 @ 5:05.
http://hps.org/hpspublications/historylnt/episodeguide.html

14. See Episode 3 @ 11:56.
http://hps.org/hpspublications/historylnt/episodeguide.html

15. See Episode 3 @ 7:35.
http://hps.org/hpspublications/historylnt/episodeguide.html

16. Mendel and His Peas—Kahn Academy
https://www.khanacademy.org/science/apbiology/heredity/mendelian-genetics-ap/a/mendel-and-his-peas
See section Scientific Legacy, para. 6.

2. McClintock and Stadler (1932–1954)
Warren Spencer (1944–1947)

1. *https://en.wikipedia.org/wiki/Barbara_McClintock*

https://www.encyclopedia.com/people/science-andtechnology/genetics-and-genetic-engineering-biographies/barbara-mcclintock
See section "Career at Cornell," para. 4.

2. What Is a Point Mutation?
https://www.genetargeting.com/mutation/what-is-a-point-mutation/

See Episode 5 @ 2:48.
http://hps.org/hpspublications/historyInt/episodeguide.html

3. See Episode 4 @ 4:20.
http://hps.org/hpspublications/historyInt/episodeguide.html

4. *https://en.wikipedia.org/wiki/Lewis Stadler*

Lewis John Stadler 1896–1954: A Biographical Memoir by M.M. Rhoades
http://www.nasonline.org/publications/biographical-memoirs/memoir-pdfs/stadler-lewis.pdf

5. Mutations in Barley Induced by X-Rays and Radium—Stadler
https://sci-hub.se/10.2307/1654017

6. Was Muller's 1946 Nobel Prize Research for Radiation-induced Gene Mutations Peer-reviewed?—Calabrese
https://peh-med.biomedcentral.com/articles/10.1186/s13010-018-0060-5

7. *https://en.wikipedia.org/wiki/James_McKeen_Cattell*

8. Muller's Nobel Prize Research and Peer Review—Calabrese (2018)
https://sci-hub.se/10.1186/s13010-018-0066-z

9. Sixth International Congress of Genetics (December 1931)
https://www.nature.com/articles/130671a0

On the Genetic Nature of Induced Mutations in Plants—Stadler
https://www.abebooks.com/Genetic-Nature-Induced-Mutations-Plants-Stadler/30671330055/bd

10. Muller and Mutations: Mouse Study of George Snell (a postdoc of Muller) fails to confirm Muller's fruit fly findings, and Muller fails to cite Snell's findings (4 April 2024)
https://link.springer.com/article/10.1007/s00204-024-03718-1

11. See Episode 5 @ 9:20.
http://hps.org/hpspublications/historylnt/episodeguide.html

12. See Episode 5 @ 14:58.
http://hps.org/hpspublications/historylnt/episodeguide.html

13. Origin of the Linearity No Threshold (LNT) Dose-Response Concept—Calabrese
https://sci-hub.se/10.1007/s00204-013-1104-7

14. Validity of the Bunsen-Roscoe Law in the Production of Mutations by Radiation of Extremely Low Intensity—Ray-Chauduri (1939)
https://era.ed.ac.uk/handle/1842/33639

Ray-Chauduri Biography:
https://www.jstor.org/stable/24104055

15. *https://en.wikipedia.org/wiki/Curt Stern*

16. Curt Stern, Who Helped Create the Modern Science of Genetics (1981)
https://www.upi.com/Archives/1981/10/26/Curt-Stern-who-helped-create-the-modern-science-of/4849372920400/

17. See Episode 7 @ 7:10.
http://hps.org/hpspublications/historylnt/episodeguide.html

18. *https://www.atomicheritage.org/profile/warren-poppino-spencer*

19. Experiments to Test the Validity of the Linear R-dose / Mutation Frequency Relation in Drosophila at Low Dosage—Spencer and Stern (1947)
https://www.ncbi.nlm.nih.gov/pmc/articles/PMC1209397/pdf/43.pdf

20. Ibid. #16 above.

21. See Episode 9 @ 2:58.
http://hps.org/hpspublications/historylnt/episodeguide.html

3. Ernst Caspari The Drosophila in the Ointment
(1945–1948)

1. See Episode 8 @ 2:08.
http://hps.org/hpspublications/historylnt/episodeguide.html

https://en.wikipedia.org/wiki/Ernst Caspari

2. Key Historical Study Findings Questioned in Debate over Threshold versus Linear Non-threshold for Cancer Risk Assessment
https://radiationeffects.org/wp-content/uploads/2022/04/Calabrese-2022_Key-historical-study-findings-questioned-in-debate-over-threshold-versus-LNT-for-cancer-risk-assess—ment.pdf

3. See Episode 6 @ 8:50.
http://hps.org/hpspublications/historylnt/episodeguide.html

Ethical Failings: The Problematic History of Cancer Risk Assessment
https://sci-hub.se/10.1016/j.envres.2020.110582
See frame 3, left column.

4. See Episode 8 @ 2:35.
http://hps.org/hpspublications/historylnt/episodeguide.html

5. See Episode 9 @ 4:25.
http://hps.org/hpspublications/historylnt/episodeguide.html

With the background rates of up to 70 mSv per year, the food grown in Kerala is ten times over the radiation limit for food in the UK. Even so, Keralans enjoy the highest life expectancy in India, with normal rates of cancer:

Brave New Climate—What We Can Learn from Kerala
https://bravenewclimate.com/2015/01/24/what-can-we-learn-from-kerala/
See para. 5.

Background Radiation and Cancer Incidence in Kerala, India-Karanagappally cohort study
https://pubmed.ncbi.nlm.nih.gov/19066487/

High Level Natural Radiation Areas in Kerala: No Evidence of Adverse Health Effects *https://www.eurasiareview.com/09042022-high-level-natural-radia-tion-areas-in-kerala-no-evidence-of-adverse-health-effects-analysis/*

Radioactivity in the Diet of Population of the Kerala Coast Including Monazite Bearing High Radiation Areas
https://journals.lww.com/healthphysics/Abstract/1970/10000/Radioac-tivity_in_the_Diet_of_Population_of_the.8.aspx

Times of India—Kerala Has the Highest Lifespan of 74.9 years
https://journals.lww.com/healthphysics/Abstract/1970/10000/Radioac-tivity_in_the_Diet_of_Population_of_the.8.aspx

6. See Episode 9 @ 4:39.
http://hps.org/hpspublications/historylnt/episodeguide.html

7. Confirmation that Hermann Muller was Dishonest in His Nobel Prize Lecture
https://pubmed.ncbi.nlm.nih.gov/37665363/

8. See Episode 8 @ 5:00.
http://hps.org/hpspublications/historylnt/episodeguide.html

9. See Episode 8 @ 11:20.
http://hps.org/hpspublications/historylnt/episodeguide.html

Also see Episode 8 @ 12:00.
http://hps.org/hpspublications/historylnt/episodeguide.html

10. Muller's Nobel Lecture on Dose-Response for Ionizing Radiation: Ideology or Science?
https://sci-hub.se/10.1007/s00204-011-0728-8

Muller's Nobel Prize data: Getting the Dose Wrong and Its Significance
https://sci-hub.se/https://pubmed.ncbi.nlm.nih.gov/31228809/

Muller's Nobel Prize Lecture: When Ideology Prevailed over Science
https://sci-hub.se/https://pubmed.ncbi.nlm.nih.gov/22166484/
11. Muller's Nobel Lecture Transcript:
https://www.nobelprize.org/prizes/medicine/1946/Muller/lecture/

Also see Episode 8 @ 13:00.
http://hps.org/hpspublications/historylnt/episodeguide.html

4. Delta Uphoff and Curt Stern (1946–1949)

1. Atomic Veterans Were Silenced for 50 years. Now They're Talking
https://www.theatlantic.com/video/index/590299/atomic-soldiers/

https://en.wikipedia.org/wiki/Atomic_veteran

2. The House the Russells built—Oak Ridge National Labs
https://www.ornl.gov/blog/ornl-review/house-russells-built

3. The Mouse House: A Brief History of the ORNL Mouse-genetics Program, 1947–2009—Liane Russell
https://www.sciencedirect.com/science/article/pii/S1383574213000690
See section 5.1. Specific-locus test (SLT). First test results published in 1951.

4. The Influence of Chronic Irradiation with Gamma Rays at Low Dosages on the Mutation Rate in Drosophila Melanogaster—Caspari and Stern (1947)
https://www.ncbi.nlm.nih.gov/pmc/articles/PMC1209399/pdf/75.pdf

5. See Episode 8 @ 19:30.
http://hps.org/hpspublications/historylnt/episodeguide.html

6. See Episode 8 @ 20:45.
http://hps.org/hpspublications/historylnt/episodeguide.html

7. See Episode 8 @ 16:55.
http://hps.org/hpspublications/historylnt/episodeguide.html

8. See Episode 9 @ 14:50.
http://hps.org/hpspublications/historylnt/episodeguide.html
9. See Episode 9 @ 15:40.
http://hps.org/hpspublications/historylnt/episodeguide.html

Key Studies Used to Support Cancer Risk Assessment Questioned
http://www.junkscience.com/wp-content/uploads/2011/09/calabrese-stern-paper.pdf
See frame 6.

10. See Episode 9 @ 17:00.
http://hps.org/hpspublications/historylnt/episodeguide.html

Influence of 24-hour Gamma-ray Irradiation at Low Dosage on the Mutation Rate in Drosophila—Uphoff and Stern (1947) [declassified] *https://cipi.com/PDF/uphoff1947.pdf*

11. Episode 10 @ 2:15
http://hps.org/hpspublications/historylnt/episodeguide.html

12. Manhattan Project Genetic Studies: Flawed Research Discredits LNT Recommendations
https://www.sciencedirect.com/science/article/abs/pii/S026974912202 1170

13. Robley D. Evans, nuclear physicist, pioneer of nuclear medicine and past president of the Health Physics Society
https://en.wikipedia.org/wiki/Robley_D._Evans_(physicist)

14. Recent Discoveries on the Historical Foundations of Cancer Risk Assessment: Shedding Light on the Limits of LNT
https://www.sciencedirect.com/science/article/abs/pii/S004896972403 8233
See page 3.

15.
https://www.sciencedirect.com/science/article/abs/pii/S000927972300 2818

Also see Episode 10 @ 4:35.
http://hps.org/hpspublications/historylnt/episodeguide.html

16. See Episode 4 @ 6:00.
http://hps.org/hpspublications/historylnt/episodeguide.html

17. See Episode 9 @ 16:30.
http://hps.org/hpspublications/historylnt/episodeguide.html

Uphoff and Stern 1947 report: Ibid. #10 above, second link

18. Cancer Risk Assessment, Its Wretched History and What It Means for Public Health
https://www.tandfonline.com/doi/full/10.1080/15459624.2024.231 1300
See Part 3, para. 4.

19. Francis Bacon (1561–1626) English philosopher, statesman, and author, is considered to be one of the founders of scientific research and the scientific method.
https://www.worldhistory.org/Scientific_Method/

20. The Genetics Effects of Low Intensity Radiation—Uphoff and Stern (1949)
https://scihub.se/https://sciub.ru/https:/www.science.org/doi/10.1126/science.109.2842.609

Also see Episode 8 @18:30 and Episode 9 @4:50.
http://hps.org/hpspublications/historylnt/episodeguide.html

21. See Episode 10 @ 4:00.
http://hps.org/hpspublications/historylnt/episodeguide.html

5. Detlev Bronk, Warren Weaver, and Jim Neel (1956)

1. Threats to Our Nation, 1957–1959: A Public Health Retrospective
https://www.ncbi.nlm.nih.gov/pmc/articles/PMC2646491/

Radioactive Fallout from Global Weapons Testing
https://www.cdc.gov/nceh/radiation/fallout/rf-gwt_home.htm

2. *https://www.britannica.com/biography/Barbara-McClintock*

3. The State of US Science and Engineering (2020 / National Science Foundation)
https://ncses.nsf.gov/pubs/nsb20201/u-s-r-d-performance-and-funding

4. The Rockefeller Foundation's Role in Creating the Atomic Bomb
https://resource.rockarch.org/story/the-atomic-bomb-development-rockefeller-foundation-role/

5. Eisenhower's "Atoms for Peace" Speech
https://ahf.nuclearmuseum.org/ahf/key-documents/eisenhowers-atoms-peace-speech/

6. Nuclear Energy and Fossil Fuels—Hubbert
http://large.stanford.edu/courses/2012/ph240/plugis2/docs/1956.pdf

7. *https://en.wikipedia.org/wiki/Detlev_Bronk*

8. *https://en.wikipedia.org/wiki/Dean_Rusk*

9. See Episode 10 @ 23:00.
http://hps.org/hpspublications/historylnt/episodeguide.html

10. See Episode 11 @ 5:40.
http://hps.org/hpspublications/historylnt/episodeguide.html

11. *https://en.wikipedia.org/wiki/Warren_Weaver*

12. *https://en.wikipedia.org/wiki/Alfred_Sturtevant*

13. Ethical Issues in the US 1956 National Academy of Sciences
BEAR-I Genetics Panel Report to the Public
*https://local.ans.org/ne/wp-content/uploads/2022/07/HP-Ethical_
Issues_in_the_US_1956_National_Academy_of.28.pdf*
See frame 4, bottom left.

14. Japanese Legacy Cohorts: The Life Span Study Atomic Bomb Survivor Cohort and Survivors' Offspring—Neel & Schull (1956)
*https://www.researchgate.net/publication/323822866_Japanese_Lega
cy_Cohorts_The_Life_Span Study Atomic Bomb Survivor Cohort and
Survivors' Offspring*

15. *https://en.wikipedia.org/wiki/William_Schull*

16. The Life-Span Study
https://www.ncbi.nlm.nih.gov/pmc/articles/PMC4674181/

17. *https://en.wikipedia.org/wiki/Radiation Effects Research Foundation*
https://www.ncbi.nlm.nih.gov/pmc/articles/PMC4674181/

18. Ibid. #13 above

19. The Muller-Neel Dispute and the Fate of Cancer Risk Assessment
https://sci-hub.se/10.1016/j.envres.2020.109961
See frame 3, left column.

20. *https://en.wikipedia.org/wiki/Tracy_Sonneborn*

21. See Episode 11 @ 12:50.
http://hps.org/hpspublications/historylnt/episodeguide.html

22. The BEAR-I Report to the Public (12 June 1956)
https://babel.hathitrust.org/cgi/pt?id=mdp.39015049805065&view=1u p&seq=8&skin=2021

23. Cancer Risk Assessment, Its Wretched History and What It Means for Public Health
https://www.tandfonline.com/doi/full/10.1080/15459624.2024.2311300
See Part 7, para. 5.

24. See Episode 13 @ 1:39.
http://hps.org/hpspublications/historylnt/episodeguide.html

25. *https://en.wikipedia.org/wiki/Arthur_Ochs_Sulzberger*

26. *https://www.youtube.com/watch?v=RallNkphNto*

https://www.youtube.com/watch?v=sU9bvag_wkQ

6. Warren Weaver and George Beadle
(1956–1957)

1. See Episode 12 @ 1:12.
http://hps.org/hpspublications/historylnt/episodeguide.html

2. See Episode 12 @ 8:40.
http://hps.org/hpspublications/historylnt/episodeguide.html

3. See Episode 12 @ 10:17.
http://hps.org/hpspublications/historylnt/episodeguide.html

4. See Episode 12 @ 11:05.
http://hps.org/hpspublications/historylnt/episodeguide.html

5. British Medical Research Council The Hazards to Man of Nuclear and Allied Radiations (June 1956)
https://cipi.com/PDF/MedicalResearchCouncil1956HazardsToMan.pdf

6. Ibid. #5 above, para. 50. Also see in the same paper:

Paragraph 63. Nevertheless, on the basis of the known levels of external radiation that can be tolerated, and of experience gained from the accidental ingestion of substances such as 'radium, permissible', levels of exposure for a large number of radioactive isotopes have been agreed. No single level that comprehends them all can be given, since each isotope emits its own characteristic amount and type of radiation, dies away at its own rate; and is absorbed and excreted at a rate dependent on its chemical form and its method of entry into the body. However, it has now been possible to estimate for many different isotopes the concentrations that it is safe to accumulate.

7. Ibid. #5 above. Also see:

Paragraph 254. Of the delayed effects, the one about which we have most information is leukaemia. it also appears to be the most easily induced and seems, at present, to be the most important as far as radiation of the whole body is concerned. We have therefore taken the incidence of leukaemia as a measure of the doses of radiation that are capable of producing delayed effects. The statistical evidence indicates that an increased incidence of leukaemia can be demonstrated after exposure to doses of radiation which might, in exceptional circumstances, be met with in civil life. For example, after either a single exposure of 200 r [200 Roentgen, or about 1,800 mSv] or a few exposures which in total amount to 200 r, there is a noteworthy increase in the small chance of developing this disease. What we do not know for certain is whether there would be an increase if a total dose of 200 r [1,800 mSv] were spread over many years. Be this as it may, however, *any risk that there may be from such a dose appears to be within the range of risks of other kinds commonly incurred in industrial and professional life.* [*emphasis added*]

8. *https://academic.oup.com/jhered/article-abstract/47/2/87/8765 45*

9. See Episode 15 @ 6:50.
http://hps.org/hpspublications/historylnt/episodeguide.html

Geneticists Find No Atomic Harm—John Hillaby, New York Times 2 August 1956
https://www.nytimes.com/1956/08/02/archives/geneticists-find-no-atomic-harm-differ-at-parley-on-ultimate.html

The Muller-Neel Dispute and the Fate of Cancer Risk Assessment
https://sci-hub.se/10.1016/j.envres.2020.109961
Search for "Copenhagen".

10. *http://calteches.library.caltech.edu/188/1/beadle.pdf*

7. Ed Lewis (1955–1958)

1. *https://en.wikipedia.org/wiki/Linus_Pauling*

2. The Rockefeller Foundation annual report 1958
https://www.rockefellerfoundation.org/wp-content/uploads/Annual-Report-1958-1.pdf

3. *https://en.wikipedia.org/wiki/Edward B. Lewis*

Edward Lewis and Radioactive Fallout: The impact of CalTech Biologists on the Debate over Nuclear Weapons Testing in the 1950s and 60s—(Caron, 2003)
https://thesis.library.caltech.edu/1190/1/LewisandFallout.pdf

4. See Episode 16 @ 10:55.
http://hps.org/hpspublications/historylnt/episodeguide.html

5. Leukemia and ionizing radiation—E.B. Lewis (1957)
https://sci-hub.se/10.1126/science.125.3255.965

6. See "Supplement: Sex and the Single Gamete," following the last chapter of this book.

7. DNA Packaging: Nucleosomes and Chromatin
https://www.nature.com/scitable/topicpage/dna-packaging-nucleosomes-and-chromatin-310/

8. *https://en.wikipedia.org/wiki/DNA_repair*

9. DNA Proofreading and Repair—Kahn Academy
https://www.khanacademy.org/science/high-school-biology/hs-molecular-genetics/hs-discovery-and-structure-of-dna/a/dna-proofreading-and-repair

10. DNA Animation—Drew Berry and Etsuko Uno (2002–2014)
https://www.youtube.com/watch?v=7Hk9jct2ozY&t=207s

11. New Study Finds that Most Cancer Mutations Are Due to Random DNA Copying 'Mistakes'
https://www.hopkinsmedicine.org/news/media/releases/new_study_finds_that_most_cancer_mutations are due to random dna copying mistakes

12. Cancer Risk Assessment, Its Wretched History, and What It Means for Public Health
https://www.tandfonline.com/doi/full/10.1080/15459624.2024.2311300
See Part 8.

13. How frequently does mitosis occur in the average human adult? On average, a human adult has 32 trillion cells:

NewScientist.com—How many cells in the human body
https://www.newscientist.com/article/2392685-we-now-know-how-many-cells-there-are-in-the-human-body

Each healthy cell makes about 10,000 repairs per day:

MIT News—A New Look at Prolonged Radiation Exposure
http://news.mit.edu/2012/prolonged-radiation-exposure-0515

Doing the Math: 32 Trillion Cells Each Performing 10,000 Repairs per Day = 320,000,000,000,000,000 (320 quadrillion) repairs per day per average human adult.

There are 86,400 seconds per day. So 320 quadrillion ÷ 86,400 = 3.7 trillion per second.

14. ScienceAlert.com—We May Have Seriously Underestimated How Hostile Conditions on Early Earth Were
https://www.sciencealert.com/conditions-on-ancient-earth-were-probably-more-hostile-than-previously-thought

8. Metastasizing Muller's Mistake (1957–1958)

1. Low-level Radiation Exposure Less Harmful to Health than Other Modern Lifestyle Risks
http://www.spacedaily.com/reports/Low_level_radiation_exposure_less_harmful_to_health_than_other_modern_lifestyle_risks_999.html

2. Ibid. #20, Chapter 7

3. The Rise of Nuclear Fear—Spencer Weart
https://www.amazon.com/RiseNuclearFearSpencerWeart/dp/0674052331

4. Immature Sperm (Spermatagonia) Have Robust Self-Repair Mechanisms.

DNA Damage in Testicular Germ Cells and Spermatozoa
https://onlinelibrary.wiley.com/doi/full/10.1111/andr.13375
See Section 1.1: "within the spermatogonial stem cell population, DNA proof reading and repair are excellent, giving the male germ line one of the lowest spontaneous mutation levels in the body."

As spermatagonia mature into spermatazoa, they grow a tail to enable mobility and turn off their self-repair mechanisms to prepare for conception.
See:

The Capacity to Repair Sperm DNA Damage in Zygotes Is Enhanced by Inhibiting wip1 Activity
https://www.frontiersin.org/articles/10.3389/fcell.2022.841327
See Intro: "unlike oocytes, [mature] sperm are incapable of repairing DNA damage. Therefore, sperm DNA damage is repaired after fertilization in zygotes using maternal DNA repair factors."

Also see:

DNA Double Strand Break Response and Limited Repair Capacity in Mouse Elongated Spermatids
https://www.ncbi.nlm.nih.gov/pmc/articles/PMC4691157/

Note: A "spermatid" is a maturing male gamete, with a half-set of DNA, that evolves into a mature spermatazoa. DNA damage in spermatazoa is repaired by the oocyte.

5. Can Oocytes Repair Fragmented DNA of Spermatozoa?
https://intapi.sciendo.com/pdf/10.2478/acb-2020-0008
See page 3, right column: "[The] fully functional repair system of oocyte DNA is able to correct persistent unrepaired damage in both [the] maternal and paternal genome."

6. Types, Causes, Detection and Repair of DNA Fragmentation in Animal and Human Sperm Cells
https://www.ncbi.nlm.nih.gov/pmc/articles/PMC3509564
See abstract.

7. *https://en.wikipedia.org/wiki/Tardigrade*

Scientists Put Tardigrade Proteins into Human Cells. Here's What Happened
https://www.sciencealert.com/scientists-put-tardigrade-proteins-into-human-cells-heres-what-happened

9. From Muller to Lewis: A Bad Idea Writ Large (1927–1957)

1. Edward Lewis and radioactive fallout
https://thesis.library.caltech.edu/1190/1/LewisandFallout.pdf
See frame 18, page 12.

2. Leukemia: A Model Metastatic Disease
https://www.ncbi.nlm.nih.gov/pmc/articles/PMC8722462/
See abstract.

When Leukemia Spreads in the Body
https://www.healthcentral.com/slideshow/leukemia-spreads-in-body
See section 3.

3. Inflammation and Cancer
https://doi.org/10.1038/nature01322

Viruses that Can Lead to Cancer
https://www.cancer.org/healthy/cancercauses/infectiousagents/infections-that-can-lead-to-cancer/viruses.html
Search for "leukemia."

4. Cancer Rate of Atomic Veterans Is Ordinary
https://www.washingtonpost.com/archive/politics/1999/10/21/cancer-rate-of-atomic-veterans-is-ordinary/1acf562c-fd9b-421a-b0e3-bc30d75c0dac/

5. Leukemia and Your Risk Factors: Is It Hereditary?
https://www.healthline.com/health/leukemia-hereditary

6. See Episode 16 @ 16:25.
http://hps.org/hpspublications/historylnt/episodeguide.html

7. See Episode 18 @ 2:35.
http://hps.org/hpspublications/historylnt/episodeguide.html

8. Linus Pauling Day-by-Day May 1957
http://scarc.library.oregonstate.edu/coll/pauling/calendar/1957/05/index.html
Search for "leukemia/"

Also see:

Earth Is a Nuclear Planet (MikeConleyAuthor.com)

9. Weak Link Found for Fallout and Leukemia
https://www.nytimes.com/1990/07/28/us/weak-link-found-for-fallout-and-leukemia.html

10. Ibid. #8 above. Search for "Schweitzer" and "Einstein"

April 29, 1962: Paulings [sic] Protest Nuclear Testing
https://www.zinnedproject.org/news/tdih/paulings-protest-nuclear-testing/

11. The Delicate Balance of Terror (Wohlstetter 1958)
https://www.rand.org/pubs/papers/P1472.html

12. Nelson Rockefeller and Civil Defense
https://www.nps.gov/articles/coldwar_civildefense_rockefellerand-civildefense.htm

Civil Defense: Against the Silent Killer
https://content.time.com/time/subscriber/article/0,33009,892725,00.html

13. Ibid. #1 above. See frame 83 / page 77.

14. *https://en.wikipedia.org/wiki/Happy_Days*

THEM! (1954)
https://www.imdb.com/title/tt0047573/

Attack of the 50 Ft Woman (1958)
https://www.imdb.com/title/tt0051380/

15. "New Frontier"—Donald Fagen (Nightfly Album)
https://www.youtube.com/watch?v=vVy0ZVQcl7E

10. The Atomic Age Hits the LNT Iceberg (1957)

1. Fallout from Nuclear Weapons Tests—Special Subcommittee
https://exhibits.stanford.edu/atomic-energy/catalog/wr029sz0509

2. Ibid. #1, Chapter 9. See frame 31 / page 25.

3. LNT and Cancer Risk Assessment: Its Flawed Foundations Part 2: How Unsound LNT Science Became Accepted
https://sci-hub.se/10.1016/j.envres.2021.111041
See introduction.

4. *https://en.wikipedia.org/wiki/Precautionary_principle*

5. Ibid. #1, Chapter 9. See frame 50 / page 44.

The Nature of Radioactive Fallout and Its Effects on Man
https://www.google.com/books/edition/The_Nature_of_Radioactive_F allout_and_It/VmtXlq_y2NgC?hl=en
See page 2,000.

6. *Washington Post Archives.* July 1957. See pages A1 and A6: Unna, W., All Radiation Held Perilous: Nation's Top Geneticists Unanimous in Opinion, Fallout Produced Now Will Shorten Lives in Future, Congress Is Told.

7. See *Life Magazine* 10 June 1957 pages 24–29.
Nuclear Worries: The Nation Begins to Worry in Earnest about Its Nuclear Tests as Scientists Explain Their Fallout Warnings to a Congressional Committee

8. No More War! Pauling, L.C., 1958. Dodd, Mead
See page 254.

9. See Episode 18 @ 7:00.
http://hps.org/hpspublications/historylnt/episodeguide.html

10. Background Material for the Development of Radiation Protection
https://www.epa.gov/sites/default/files/201505/documents/frc_rpt1.pdf
See frame 27 / page 23.
(This is a 1960 publication of the Federal Radiation Council, which quotes excerpts from the NCRP's Handbook 59, published in 1954.)

11. Edward B. Lewis, 1918—2004: A Hagiography by Crow and Bender
ncbi.nlm.nih.gov/pmc/articles/PMC1448758/
See para. 5 of the section "Radiation and Cancer":

"For genetic effects, the linear, non-threshold assumption aroused little controversy, partly because of the eminence of Sturtevant and Muller, although the estimated numbers were in considerable doubt. The reaction to somatic effects was quite different and Ed Lewis quickly found himself embroiled in controversy. His work was challenged by Neil Wald from the Atomic Bomb Casualty Commission in Japan, by Austin Brues of the Argonne Laboratories (Brues 1958), and by no less than Admiral Lewis L. Strauss, Chairman of the Atomic Energy Commission (AEC; Lipshitz 2004a, page 396). Ed's critics also included Caltech faculty and administrators. The most detailed criticism came from A.W. Kimball, a statistician at the Oak Ridge Laboratory."

12. See Episode 18 @ 6:28.
http://hps.org/hpspublications/historylnt/episodeguide.html

13. See Episode 18 @ 7:20.
http://hps.org/hpspublications/historylnt/episodeguide.html

14. Radiation Dose Rate and Mutation Frequency—Russell (1958)
https://sci-hub.ru/https:/pubmed.ncbi.nlm.nih.gov/13615306/

Also see:

Episode 20 @ 11:00
http://hps.org/hpspublications/historylnt/episodeguide.html
27,000 x 3 mSv (average annual background radiation on Earth) equals about 80,000 mSv.
Also see endnote #2 to next chapter.

15. ALARA—As Low As Reasonably Achievable
https://www.cdc.gov/nceh/radiation/alara.html

ALARA—The Gold Standard of Radiation Protection
https://www.versantphysics.com/2021/04/08/alara-the-gold-standard-of-radiation-protection/

16. LNT and Cancer Risk Assessment: Its Flawed Foundations Part 1. Radiation and Leukemia, Where LNT Began
https://sci-hub.se/10.1016/j.envres.2021.111025
See frame 12 / page 12 right column #11.

17. See Episode 18 @ 3:50.
http://hps.org/hpspublications/historylnt/episodeguide.html

11 William and Liane Russell and Paul Selby
(1947–1996)

1. "A Life Well Lived:" Symposium Honors Research Legacy of Liane Russell
https://www.ornl.gov/news/life-well-lived-symposium-honors-research-legacy-liane-russell

2. See Episode 20 @ 10:53.
http://hps.org/hpspublications/historylnt/episodeguide.html

The Threshold vs LNT Showdown: Dose Rate Findings Exposed Flaws in the LNT Model, Part 1. The Russell-Muller Debate
https://sci-hub.se/10.1016/j.envres.2016.12.006
Section 2.2

Note: The average background dose-rate ranges from about 1.5 mSv to about 3.5 mSv per year, depending on where you are:

World-Nuclear.org—What Is Radiation?
https://www.world-nuclear.org/uploadedFiles/org/Features/Radiation/4Background Radiation%281%29.pdf

Russell assumed a yearly background dose-rate of 1.75 mSv. At that rate, the 0.09 mSv doses he administered were 27,000 times background. Assuming 3 mSv per year, the female mice exhibited no genetic damage at 16,000 times background dose.

3. See Episode 20 @ 12:45.
http://hps.org/hpspublications/historylnt/episodeguide.html

4. See Episode 20 @ 15:35.
http://hps.org/hpspublications/historylnt/episodeguide.html

6. See Episode 20 @ 10:00.
http://hps.org/hpspublications/historylnt/episodeguide.html

7. See Episode 20 @ 3:38.
http://hps.org/hpspublications/historylnt/episodeguide.html

8. See Episode 21 @ 8:00.
http://hps.org/hpspublications/historylnt/episodeguide.html

9. Effects on Populations of Exposure to Low Levels of Ionizing Radiation (1972)
https://nap.nationalacademies.org/catalog/18994/effects-on-populations-of-exposure-to-low-levels-of-ionizing-radiation

10. EPA adopts LNT: New Historical Perspectives
https://scihub.se/https://www.sciencedirect.com/science/article/abs/pi i/S0009279719307033

Radiation Health Effects (EPA)
https://www.epa.gov/radiation/radiation-health-effects

11. Three Mile Island Accident (World Nuclear Association)
https://www.world-nuclear.org/information-library/safety-and-secu-rity/safety-of-plants/three-mile-island-accident.aspx

"Indeed, more than a dozen major, independent health studies of the accident showed no evidence of any abnormal number of cancers around TMI years after the accident. The only detectable effect was psychological stress during and shortly after the accident."

"The studies found that the radiation releases during the accident were minimal, well below any levels that have been associated with health effects from radiation exposure. The average radiation dose to people living within 10 miles of the plant was 0.08 millisieverts, with no more than 1 millisievert to any single individual. *The level of 0.08 mSv is about equal to a chest X-ray, and 1 mSv is about a third of the average background level of radiation received by US residents in a year.*" [*emphasis added*]

12. See Episode 21@ 4:50.
http://hps.org/hpspublications/historylnt/episodeguide.html

13. See Episode 21 @ 7:20.
http://hps.org/hpspublications/historylnt/episodeguide.html

14. The Selby-Russell Dispute Regarding the Nonreporting of Critical Data in the Mega-Mouse Experiments of Drs William and Liane Rus-sell that Spanned Many Decades: What Happened, Current Status, and Some Ramifications (12 February 2020 / Selby)
https://www.ncbi.nlm.nih.gov/pmc/articles/PMC7016328/

15. Spontaneous Mutations Recovered as Mosaics in the Mouse spe-cific-locus test (Russell and Russell)
https://www.ncbi.nlm.nih.gov/pmc/articles/PMC24048/

Also see Episode 21 @ 17:58.
http://hps.org/hpspublications/historylnt/episodeguide.html

16. See Episode 21 @ 18:20.
http://hps.org/hpspublications/historylnt/episodeguide.html

17. Cover Up and Cancer Risk Assessment: Prominent US Scientists Suppressed Evidence to Promote Adoption of LNT
https://tinyurl.com/3e5xxssy

18. radiationeffects.org – What's Wrong With Being Cautious?
https://radiationeffects.org/wp-content/uploads/2015/05/Rockwell-NN-1997_Whats-wrong-with-being-cautious.pdf

19. See Episode 21 @ 20:55.
http://hps.org/hpspublications/historylnt/episodeguide.html

The Linear No-Threshold (LNT) Dose Response Model: A Comprehensive Assessment of Its Historical and Scientific Foundations
https://www.sciencedirect.com/science/article/pii/S0009279718311177
See section 11, para. 1.

20. Recent Discoveries on the Historical Foundations of Cancer Risk Assessment: Shedding Light on the Limits of LNT
https://www.sciencedirect.com/science/article/abs/pii/S0048969724038233
See page 5.

The Gofman-Tamplin Cancer Risk Controversy and Its Impact on the Creation of BEIR I and the Acceptance of LNT
https://www.ncbi.nlm.nih.gov/pmc/articles/PMC9987470
See abstract.

23. World Nuclear.org—Fukushima Radiation Exposure
https://www.world-nuclear.org/information-library/safety-and-security/safety-of-plants/appendices/fukushima-radiation-exposure.aspx
See section "Radiation Effects," para.1.

Also see: *Earth Is a Nuclear Planet*, Chapter 1
https://www.MikeConleyAuthor.com

Radiation Doses from CT Scans
www.webmd.com/cancer/radiation-doses-ct-scans

24. US Coal Power Plants Killed at Least 460,000 People in Past 20 years—Report
https://www.theguardian.com/environment/2023/nov/23/coal-power-plants-deaths-pollution

25. Leukemia and Ionizing Radiation Revisited (2015 Cuttler, Welch)
https://www.omicsonline.org/open-access/leukemia-and-ionizing-radiation-revisited-2329-6917-1000202.pdf

Evidence of a Dose Threshold for Radiation-Induced Leukemia (Cuttler)
https://www.ncbi.nlm.nih.gov/pmc/articles/PMC6247492/

Leukemia Incidence of 96,000 Hiroshima Atomic Bomb Survivors Is Compelling Evidence that the LNT Model Is Wrong (Cuttler)
https://pubmed.ncbi.nlm.nih.gov/24504164/

26. Cuttler and Welsh Revisit Lewis (2015)
2. Ed Calabrese—An introduction to Hormesis and How It Can Affect Our Lifespan
https://www.youtube.com/watch?v=Vxbe1MflJ_A

12. Cutler and Welsh Revisit Lewis (2015)

1. Leukemia and Ionizing Radiation Revisited (2015, Cuttler, Welch)
https://www.researchgate.net/publication/338819152_Leukemia_and_Ionizing_Radiation_Revisited

2. Ed Calabrese—An introduction to Hormesis and How It Can Affect Our Lifespan
https://www.youtube.com/watch?v=Vxbe1MflJ_A

3. Historical blunders: How Toxicology Got the Dose-Response Relationship Half Right
https://tinyurl.com/bdfkb76z

Hormesis and Homeopathy—the Artificial Twins
https://www.ncbi.nlm.nih.gov/pmc/articles/PMC4566758

4. *https://en.wikipedia.org/wiki/Homeopathic_dilutions*

5. Ibid. #1 above.

6. See Episode 1 @ 1:10.
http://hps.org/hpspublications/historylnt/episodeguide.html

7. Cancer Stat Facts: Leukemia
https://seer.cancer.gov/statfacts/html/leuks.html

8. See Episode 17 @ 19:00.
http://hps.org/hpspublications/historylnt/episodeguide.html

9. Infection as a Cause of Childhood Leukemia: Virus Detection Employing Whole Genome Sequencing
https://www.ncbi.nlm.nih.gov/pmc/articles/PMC5685284/

10. Forbes – After five years, what is the cost of Fukushima?
https://www.forbes.com/sites/jamesconca/2016/03/10/after-five-years-what-is-the-cost-of-fukushima/
See para. 3

JCER.or.jp – Accident Cleanup Costs Rising to 35–80 Trillion Yen in 40 Years
https://www.jcer.or.jp/jcer_download_log.php

11. Ibid. #11 above, second link. See first two pages. Also see:

Reuters – Japan panel recommends ocean release for contaminated Fukushima water
https://www.reuters.com/article/world/japan-panel-recommends-ocean-release-for-contaminated-fukushima-water-idUSKBN1ZU25Z/

World Nuclear.org – Discharge to sea will have minimal impact, TEPCO says
https://www.world-nuclear-news.org/Articles/Discharge-to-sea-will-have-minimal-impact-Tepco-sa

Also see the Generation Atomic calculator (1.6 MBq per liter):

Glide.page – Radiation Dose Calculator
https://radiation-dose-calculator.glide.page/dl/welcome

12. NRC.org – NRC Decommissioning Rule from 10 CFR 20 NRC radiation standards: Subpart E-Radiological Criteria for License Termination
https://www.nrc.gov/docs/ML0824/ML082480281.pdf

See pgs. 1-2: "A site will be considered acceptable for unrestricted use if the residual radioactivity that is distinguishable from background radiation results in a TEDE to an average member of the critical group that does not exceed 25 mrem (0.25 mSv) per year, including that from groundwater sources of drinking water, and the residual radioactivity has been reduced to levels that are as low as reasonably achievable (ALARA)."

13. NRC.gov—Assessment of Variations of Radiation Exposure in the United States

https://www.nrc.gov/docs/ML1224/ML12240A227.pdf
See frame 6 / pg. 4

14. Leukemia and Nutrition: Malnutrition Is an Adverse Prognostic Factor
https://pubmed.ncbi.nlm.nih.gov/2586144/

13. Calabrese, Cuttler, and McNutt (2015–present)

1. See Episode 17 @ 2:41.
http://hps.org/hpspublications/historylnt/episodeguide.html

2. See Episode 17 @ 17:30.
http://hps.org/hpspublications/historylnt/episodeguide.html

3. See Episode 4 @ 6:00.
http://hps.org/hpspublications/historylnt/episodeguide.html

4. Societal Threats from Ideologically-Driven Science
https://www.nas.org/academic-questions/30/4/societalthreats from_ideologically_driven_science/pdf

5. See Episode 22 @ 1:35.
http://hps.org/hpspublications/historylnt/episodeguide.html
6. Leveraging Advances in Modern Science to Revitalize Low-Dose Radiation Research in the United States (2022) [NASEM proposal]
https://nap.nationalacademies.org/catalog/26434/leveraging-advances-in-modern-science-to-revitalize-low-dose-radiation-research-in-the-united-states

7. World Nuclear.org—World Nuclear Performance Report 2019
https://world-nuclear.org/images/articles/performance-report-2019.pdf

See Section 2.1 Global Highlights, page 12 figure 1. By scaling of the "Nuclear Electricity Production" bar graph, the purple segments for North America show that total historical US and Canadian nuclear electric production is about 30,000 TWh, or 30 trillion kWh.

Working with the historical Canada reactor chart and other data in this link indicates that total historical Canadian nuclear electric production is about 3.3 trillion kWh. Therefore, the US portion of the historical total is about 27 trillion kWh.

8. The world has about 1,600 billion barrels (1.6 trillion) of recoverable reserves:
Rystadenergy.com – Recoverable Oil Reserves
https://www.rystadenergy.com/news/recoverable-oil-reserves-billion-barrels-warming-planet

At a conservative price of $60 a barrel, that's $96 trillion:
macrotrends.net – Crude Oil History Chart
https://www.macrotrends.net/1369/crude-oil-price-history-chart

9. Ibid. #6 above. See Preface, page xiii.

10. *https://www.rollingstone.com/politics/politics-news/global-warmings-terrifying-new-math-188550/*

11. Energy.gov—Five Radioactive Products We Use Every Day
https://www.energy.gov/ne/articles/5-radioactive-products-we-use-every-day

NRC.gov—Assessment of Variations of Radiation Exposure in the United States
https://www.nrc.gov/docs/ML1224/ML12240A227.pdf

14. A Way Forward

1. See Episode 22 @ 16:15.
http://hps.org/hpspublications/historylnt/episodeguide.html

2. Wiki—List of Causes by Death Rate
https://en.wikipedia.org/wiki/List_of_causes_of_death_by_rate

See table: Leading Causes of Death by Age Groups, US—2015. Note that in age group 45–54, 43,000 cancer deaths out of 136,000 deaths (top ten causes) = 32 percent. In age group 55–64, 116,000 cancer deaths out of 288,000 deaths = 40 percent. In age group 65-plus, 596,000 cancer deaths out of 1,514,000 deaths = 39 percent.

22% of deaths in US, 2017:

What Are the 10 Leading Causes of Death in the United States?
https://www.healthline.com/health/leading-causes-of-death

16 percent globally in 2016:

What do the people of the world die from? (BBC)
https://www.bbc.com/news/health-47371078

Global Cancer Facts and Figures, fourth edition
https://www.cancer.org/content/dam/cancer-org/research/cancer-facts-and-statistics/global-cancer-facts-and-figures/global-cancer-facts-and-figures-4th-edition.pdf

3. See Episode 22 @ 17:55.
http://hps.org/hpspublications/historylnt/episodeguide.html

4. See Episode 22 @ 19:10.
http://hps.org/hpspublications/historylnt/episodeguide.html

Also see:

See Episode 22 @ 17:00 (Cancer Risk Model Optimization)
http://hps.org/hpspublications/historylnt/episodeguide.html

The EPA Cancer Risk Assessment Default Model Proposal: Moving Away from the LNT
https://sci-hub.se/10.1177/1559325818789840

Integration of Hormesis and LNT Optimizes Cancer Risk Assessment (Calabrese slide deck / presentation)
https://www.toxicology.org/groups/ss/RSESS/doc/RASS-RSESS_Webinar_101415.pdf

The Integration of LNT and Hormesis for Cancer Risk Assessment Optimizes Public Health protection
https://sci-hub.se/10.1097/hp.0000000000000382

Supplement: Sex and the Single Gamete

1. Mitosis Observed—National Human Genome Research Institute
https://www.genome.gov/25520234/online-education-kit-1879-mitosis-observed
https://www.news-medical.net/life-sciences/Mitosis-vs-Meiosis.aspx

2. The mechanism of DNA replication was initially described in the 1960s:

Biologic Synthesis of Deoxyribonucleic Acid
https://www.science.org/doi/10.1126/science.131.3412.1503

The mechanisms of mitosis and meiosis weren't worked out until the 1990s:

Meiosis: From Molecular Basis to Medicine
https://www.ncbi.nlm.nih.gov/pmc/articles/PMC8671932/

3. Sorcerer's Apprentice sequence in Fantasia
https://video.disney.com/watch/sorcerer-s-apprentice-fantasia-4ea9ebc01a74ea59a5867853

4. Mitosis—Kahn Academy
https://www.khanacademy.org/science/ap-biology/cell-communication-and-cell-cycle/cell-cycle/a/phases-of-mitosis

5. DNA Proofreading and Repair—Kahn Academy
https://www.khanacademy.org/science/high-school-biology/hs-molecular-genetics/hs-discovery-and-structure-of-dna/a/dna-proofreading-and-repair

6. Centromere
https://www.genome.gov/genetics-glossary/Centromere

7. Sexual Life Cycles—Kahn Academy
https://www.khanacademy.org/science/biology/cellular-molecular-biology/meiosis/a/sexual-life-cycles

8. How Many Eggs Does a Woman Have?
https://www.evewell.com/support/how-many-eggs-does-a-woman-have/

9. Meiosis PMAT Phase I—Kahn Academy
https://www.khanacademy.org/science/ap-biology/heredity/meiosis-and-genetic-diversity/v/phases-of-meiosis-i

Meiosis PMAT Phase II—Kahn Academy
https://www.khanacademy.org/science/ap-biology/heredity/meiosis-and-genetic-diversity/v/phases-of-meiosis-ii

Now That You've Read This Book . . .

Suggestions for Further Reading

The controversial subject of LNT has been mostly explored in scientific studies and technical papers, many of which are referenced in the Endnotes to this book. Here are several outstanding books, and a couple of outstanding articles, which are all technically sound yet highly readable.

John D. Boice, Jr, Andre Bouville, Lawrence T. Dauer, Ashley P. Golden, and Richard Wakeford, eds. 2025. *The Million Person Study of Low-Dose Radiation Health Effects*. CRC Press.

Edward J. Calabrese, ed. 2018 [1999]. *Biological Effects of Low Level Exposures to Chemicals and Radiation: Dose-Response Relationships*. CRC Press.

Mike Conley and Tim Maloney. 2024. *Earth Is a Nuclear Planet: The Environmental Case for Nuclear Power*. Carus Books.

Sohei Kondo. 1993. *Health Effects of Low-Level Radiation*. Medical Physics Publishing.

J.M. Cuttler. 2007. Health Effects of Low Level Radiation: When Will We Acknowledge the Reality? *Dose Response* 5:4 (September).

T.D. Luckey. 1991. *Radiation Hormesis*. CRC Press.

Mark P. Mattson and Edward J. Calabrese, eds. 2010. *Hormesis: A Revolution in Biology, Toxicology, and Medicine.* Springer.

Kate-Louise D. Gottfried and Gary Penn. 1996. *Radiation in Medicine: A Need for Regulatory Reform.* National Academy Press.

George Parris. 2019. *The Myth of the Linear, No-Threshold Dose-Response Relationship for Carcinogens.* Independently published.

Prekeges, Jennifer L. 2003. Radiation Hormesis, or, Could All That Radiation Be Good for Us? *Journal of Nuclear Medicine Technology* 31:1 (March).

Bill Sacks and Greg Meyerson. 2025. *The Poverty of Green Philosophy: A Marxist Case for Nuclear Energy in a Cooperative World.* Carus Books.

Charles L. Sanders. 2009. *Radiation Hormesis and the Linear-No-Threshold Assumption.* Springer.

Charles L. Sanders. 2007. *Radiobiology and Radiation Hormesis: New Evidence and Its Implications for Medicine and Society.* Springer.

Euclid Seeram. 2019 [2001]. *Rad Tech's Guide to Radiation Protection.* Wiley Blackwell.

Index

EARTH
— IS A —
NUCLEAR PLANET
THE ENVIRONMENTAL CASE FOR NUCLEAR POWER

MIKE CONLEY & TIM MALONEY

Senior Science Advisor:
STEPHEN A. BOYD, Ph.D.

Foreword by:
ERIC MEYER, FOUNDER OF GENERATION ATOMIC